EXAMINATION TEST PAPERS C3/C

CONTENTS

There are 10 (1 hour and 30 minutes) C3 Test Papers
10 (1 hour and 30 minutes) C4 Test Papers

	C3	C4
Test Paper 1	2–13	113–121
Test Paper 2	14–25	122–130
Test Paper 3	26–36	131–139
Test Paper 4	37–47	140–149
Test Paper 5	48–57	150–157
Test Paper 6	58–69	158–166
Test Paper 7	70–80	167–175
Test Paper 8	81–90	176–184
Test Paper 9	91–100	185–195
Test Paper 10	101–112	196–205

H. Nicolaides
08-10-2016

GCE Examinations

Test Paper 1

Advanced Level

Core Mathematics C3

Time: 1 hour 30 minutes

Instructions and Information

Candidates may use any calculator allowed by the regulations of their Examination Board.

Full marks are awarded for correct answers to ALL questions.

This paper has eight questions.

You can start working with any question and you must label clearly all parts.

1. (a) Draw a mapping diagram for the function
   $f(x) = 2x - 1 \quad -2 \leq x \leq 2$. (2)
   (b) Sketch the graph of f over its domain and state the range. (2)
   (c) Find $f^{-1}(x)$ and state the domain of f^{-1}. (4)
   (d) State the relationship between $f(x)$ and $f^{-1}(x)$. (2)

2. Simplify
   (i) $\dfrac{x+6}{x^2 + 10x + 24}$. (2)

   Show that
   (ii) $\dfrac{x^3 + 1}{x + 1} = x^2 - x + 1$. (3)

3. A right circular cone is made up of two parts, the curved part and a circle which is the base of radius r. The curved part is a sector of radius R, whose angle at the centre is ϕ as shown in Fig. 1

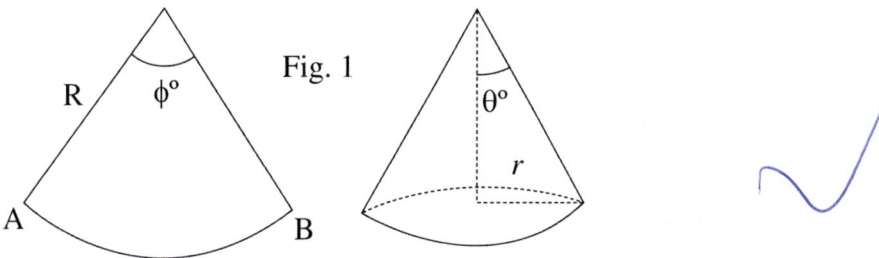

Fig. 1

Show that the area of the sector of $A = \frac{1}{2} R^2 \phi$ and the circumference of the base $AB = R\phi = 2\pi r$. (2)

The semi-vertical angle of the cone is θ radians.

Express ϕ in terms of $\sin \theta$ and show that the volume V of the cone is given by the expression
$$3V = \pi R^3 \sin^2 \theta \cos \theta$$ (6)

If R is constant and θ varies find the positive value of $\tan \theta$ for which $\frac{dV}{d\theta} = 0$.

Show further that when this value of $\tan \theta$ is taken, the maximum value V is obtained which is equal to $\dfrac{2\pi R^2 \sqrt{3}}{27}$ and find, in terms of R, the area of the sector of Fig. 1 (8)

4. (a) Given that $a \sin x + b \cos x \equiv 13 \sin \left(x + \frac{\pi}{6}\right)$ find the constants a and b. (2)

   (b) Given that $a \cos x - b \sin x \equiv 15 \cos \left(x + \frac{\pi}{4}\right)$ find the constants a and b. (2)

   (c) Solve

   (i) $13 \sin \left(x + \dfrac{\pi}{6}\right) = 5$ (3)

   (i) $15 \cos \left(x + \dfrac{\pi}{4}\right) = 12$ (3)

   where $0 \leq x \leq 2\pi$.

5. An approximate solutions of the equation
$$\sin x + \sin \frac{x}{2} - 1 = 0 \quad \text{is } x = 43.5°.$$

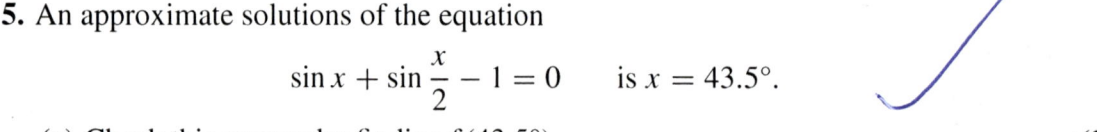

   (a) Check this answer by finding $f(43.5°)$. (2)

   (b) Which is an improved solution $40.7°$ or $45°$? Show the working. (4)

6. (a) Sketch the functions:

   (i) $\sin x$ (ii) $\cos x$ (iii) $\tan x$

   for the domain $-360° \leq x \leq 360°$. (4)

   (b) Sketch the corresponding inverse functions for (i), (ii) and (iii) for this domain.

   Label clearly the ranges and domains for all six graphs. (6)

7. The curve C with equation $y = 3e^x + 4$ intersects the y-axis at A.

   (a) Find the equations of the tangent and normal at A. (4)

   (b) Determine the coordinates of the points of intersections of the tangent and normal with the x-axis (2)

   (c) Sketch the graphs and indicate the coordinates. (2)

8. Differentiate with respect to t the following:

   (i) $e^{-t} \sin t$ (3)

   (ii) $\cos(2t^3 - 1)$ (3)

   (iii) $\dfrac{2t - 1}{t^2 + 1}$. (4)

TOTAL FOR PAPER: 75 MARKS

GCE Examinations

Test Paper 1 Solutions

Advanced Level

Core Mathematics C3

1. (a)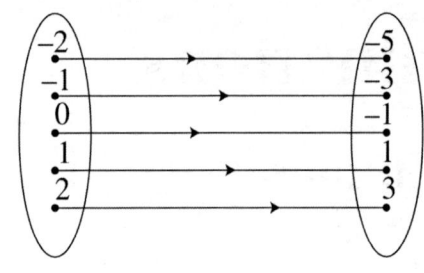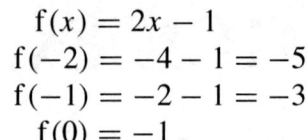

$f(x) = 2x - 1$
$f(-2) = -4 - 1 = -5$
$f(-1) = -2 - 1 = -3$
$f(0) = -1$
$f(1) = 1$
$f(2) = 3$

Domain   Co domain or Range

(b)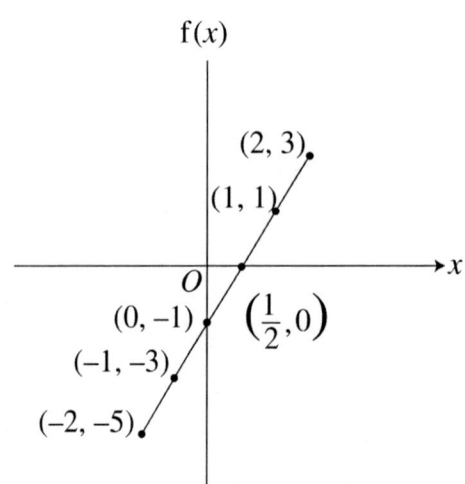

The range is $-5 \leq y \leq 3$.

(c) $y = 2x - 1$ change x for y and y for x and solve for y

$x = 2y - 1 \Rightarrow 2y = x + 1 \Rightarrow y = \frac{1}{2}x + \frac{1}{2}$

$\therefore f^{-1}(x) = \frac{1}{2}x + \frac{1}{2}$

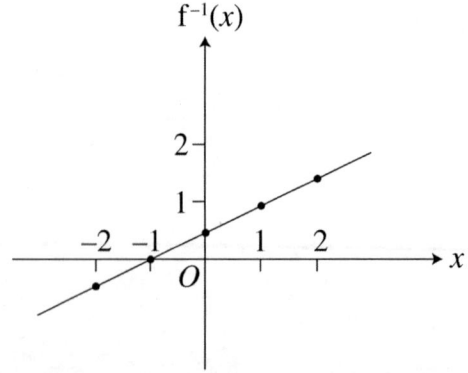

2. (i) $\dfrac{x+6}{x^2+10x+24} = \dfrac{x+6}{(x+6)(x+4)} = \dfrac{1}{x+4}$

(ii)
$$\begin{array}{r}
x^2 - x + 1 \\
x+1 \overline{\smash{)}\, x^3 + 1 } \\
\underline{x^3 + x^2} \\
-x^2 + 1 \\
\underline{-x^2 - x} \\
x + 1 \\
\underline{x+1} \\
0
\end{array}$$

$\dfrac{x^3+1}{x+1} = x^2 - x + 1.$

---

3. $A = \pi R^2 \dfrac{\phi}{2\pi} = \dfrac{1}{2} R^2 \phi$ and $AB = 2\pi R \dfrac{\phi}{2\pi} = R\phi$

hence $R\phi = 2\pi r \Rightarrow \boxed{r = \dfrac{R\phi}{2\pi}}$

$\sin\theta = \dfrac{r}{R}$

$\cos\theta = \dfrac{h}{R}$

$\sin\theta = \dfrac{r}{R} = \dfrac{R\phi}{2\pi R} = \dfrac{\phi}{2\pi} \Rightarrow \boxed{\phi = 2\pi \sin\theta}$

The volume of the cone, $V = \dfrac{1}{3}\pi r^2 h$

$= \dfrac{1}{3}\pi \dfrac{R^2 \phi^2}{4\pi^2} h$

$= \dfrac{1}{3}\pi \dfrac{R^2 \phi^2}{4\pi^2} \cos\theta\, R$

$3V = \dfrac{R^2}{4\pi} 4\pi^2 \sin^2\theta \cos\theta\, R$

$\boxed{3V = \pi R^3 \sin^2\theta \cos\theta}$

$$3\frac{dV}{d\theta} = \pi R^3 \left[2\sin\theta\cos^2\theta + \sin^2\theta(-\sin\theta)\right]$$

$$\frac{dV}{d\theta} = 0 \quad 2\sin\theta\cos^2\theta = \sin^3\theta$$

$$\sin\theta(2\cos^2\theta - \sin^2\theta) = 0$$

for maximum V

$$2\cos^2\theta - \sin^2\theta = 0$$
$$\tan^2\theta = 2$$
$$\tan\theta = \pm\sqrt{2}$$
$$\boxed{\therefore \tan\theta = \sqrt{2}}$$

$$3V = \pi R^3 \sin^2\theta \cos\theta = \pi R^3 \frac{2}{3}\frac{1}{\sqrt{3}}\frac{\sqrt{3}}{3\sqrt{3}}$$

$$V_{max} = \frac{\pi R^3\, 2\sqrt{3}}{27}.$$

The area of sector $A = \dfrac{1}{2}R^2\phi = \dfrac{1}{2}R^2 2\pi \sin\theta$

$$\pi R^2 \sin\theta = \pi R^2 \sqrt{\frac{2}{3}}$$

$$A = \pi R^2 \sqrt{\frac{2}{3}}.$$

---

4. (a) $a\sin x + b\cos x \equiv 13\sin\left(x + \dfrac{\pi}{6}\right)$

$$\equiv 13\sin x \cos\frac{\pi}{6} + 13\cos x \sin\frac{\pi}{6}$$

$$\equiv 13\frac{\sqrt{3}}{2}\sin x + 13\frac{1}{2}\cos x$$

equating coefficicents of $\sin x$ and $\cos x$ we have

$$a = \frac{13\sqrt{3}}{2} \qquad b = \frac{13}{2}$$

(b) $a\cos x - b\sin x = 15\cos\left(x + \dfrac{\pi}{4}\right)$

$$= 15\cos x \cos\frac{\pi}{4} - 15\sin x \sin\frac{\pi}{4}$$

equating coefficients of sin x and cos x
we have

$$a = 15\cos\frac{\pi}{4} = \frac{15}{\sqrt{2}} \qquad b = 15\sin\frac{\pi}{4} = \frac{15}{\sqrt{2}}$$

(c) (i) $\sin\left(x + \frac{\pi}{6}\right) = \frac{5}{13} = \sin 22.61986495°$

$$= \sin 0.394791115^c$$

$$x = 0.394791115^c - \frac{\pi}{6}$$

$$= -0.128807655$$

$$= -0.129^c \text{ to 3 d.p.}$$

(ii) $\cos\left(x + \frac{\pi}{4}\right) = \frac{12}{15} = \frac{4}{5} = 0.8 = \cos 0.643501108^c$

$$x + \frac{\pi}{4} = 0.643501108^c$$

$$x = -0.141897054^c$$

$$x = -0.142^c \text{ to 3 d.p.}$$

---

5. $f(x) = \sin x + \sin\dfrac{x}{2} - 1$

$f(43.5°) = \sin 43.5° + \sin\dfrac{43.5°}{2} - 1$

$f(43.5°) = 0.058912013$

$f(40.7°) = \sin(40.7°) + \sin 20.35° - 1 = -0.00147614308$

$f(45°) = 0.089790213$

A better approximation is $x = 40.7°$

∴ change of sign.

The actual solution $40.7° < x < 43.5°$

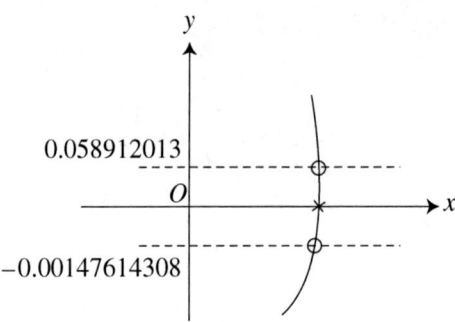

---

6. $x = \sin y \Rightarrow y = \sin^{-1} x$ so if $f(x) = \sin x$

$f^{-1}(x) = \sin^{-1} x$, similarly if $g(x) = \cos x$,

then $g^{-1}(x) = \cos^{-1} x$ and $h(x) = \tan x$, then $h^{-1} x = \tan^{-1} x$.

The x-axis represents angles and y-axis represents numbers.

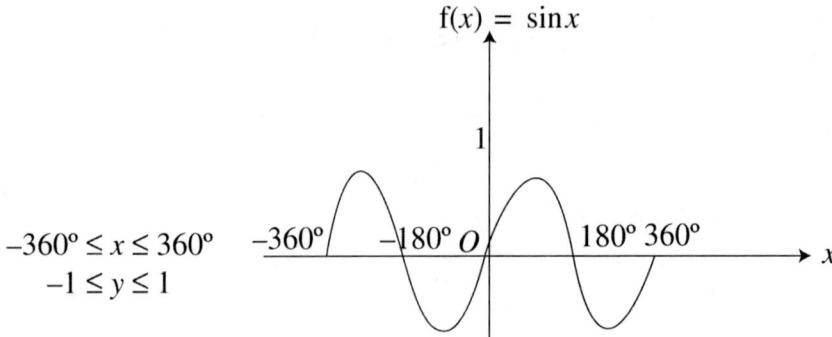

The x-axis represents numbers and y-axis represents angles.

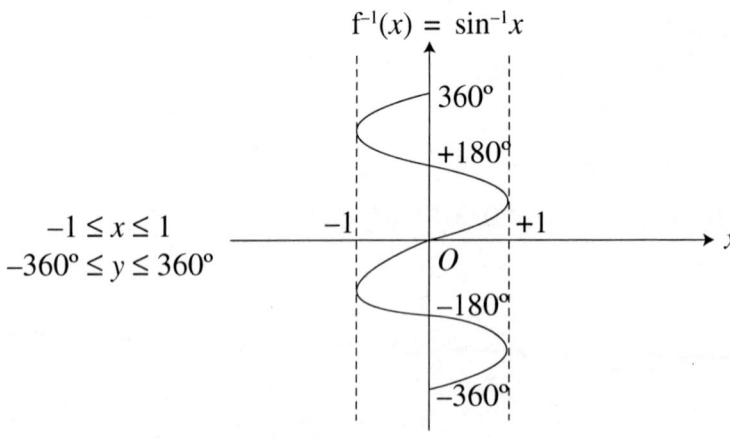

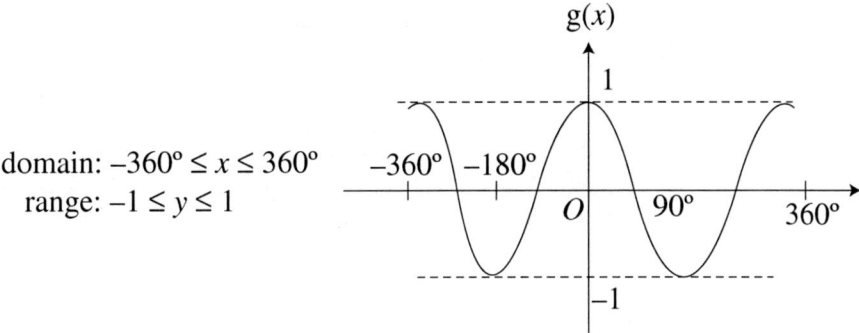

domain: $-360° \leq x \leq 360°$
range: $-1 \leq y \leq 1$

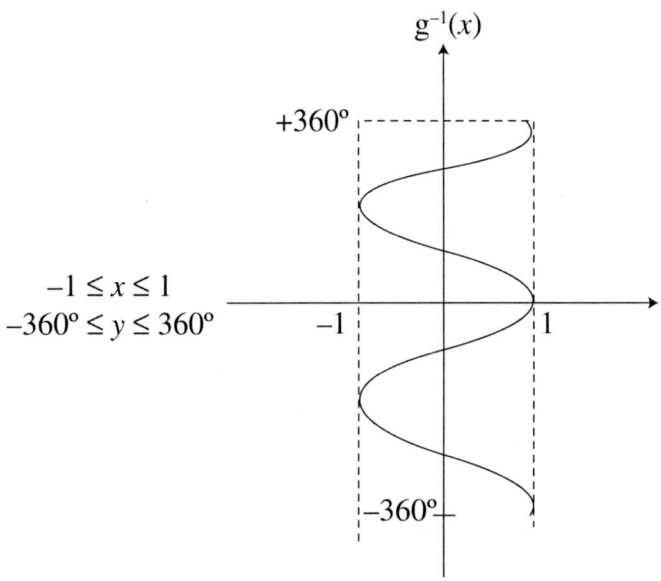

$-1 \leq x \leq 1$
$-360° \leq y \leq 360°$

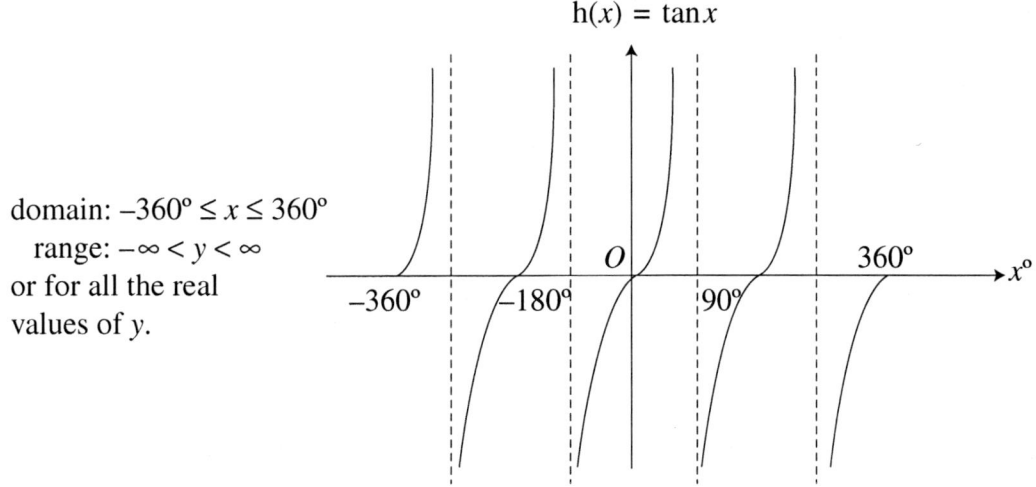

domain: $-360° \leq x \leq 360°$
range: $-\infty < y < \infty$
or for all the real values of y.

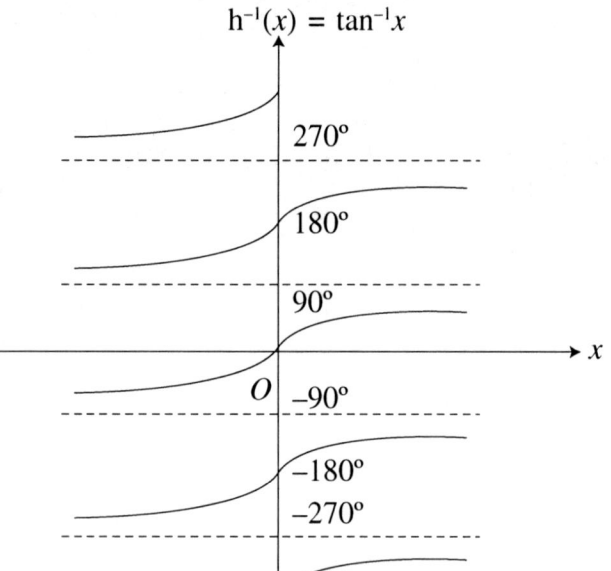

domain: for all the real values of x

range: $-360° \le y \le 360°$

7.
$$y = 3e^x + 4$$

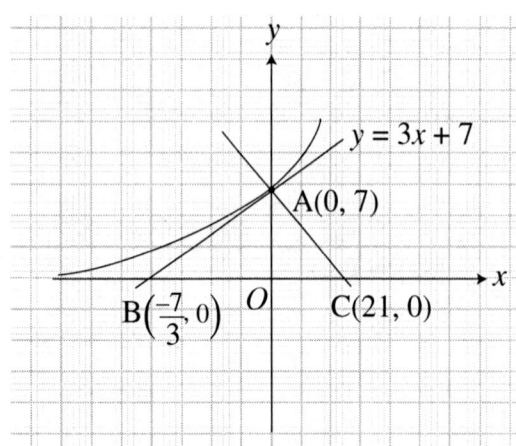

(a) $f(x) = 3e^x + 4$

$f(0) = 3 + 4 = 7$

$\therefore A(0, 7)$

$\dfrac{dy}{dx} = 3e^x$ at $x = 0$ $\quad \dfrac{dy}{dx} = 3 = m_1$ the gradient of the tangent

$y = 3x + c \Rightarrow \boxed{y = 3x + 7}$

$$m_1 m_2 = -1 \Rightarrow m_2 = -\frac{1}{3}$$

$$y = -\frac{1}{3}x + c$$

$$\boxed{y = -\frac{1}{3}x + 7}$$

(b) $y = 3x + 7$ when $y = 0$, $x = -\frac{7}{3}$

$y = -\frac{1}{3}x + 7$ when $y = 0$, $x = 21$

$B\left(-\frac{7}{3}, 0\right)$ $C(21, 0)$.

---

8. (i) $f(t) = e^{-t} \sin t$

$f'(t) = -e^{-t} \sin t + e^{-t} \cos t$
$= e^{-t}(\cos t - \sin t)$

(ii) $g(t) = \cos(2t^3 - 1)$

$g'(t) = -\sin(2t^3 - 1)6t^2$
$= -6t^2 \sin(2t^3 - 1)$

(iii) $h(t) = \dfrac{2t - 1}{t^2 + 1}$

$h'(t) = \dfrac{2(t^2 + 1) - (2t - 1)2t}{(t^2 + 1)^2}$

$= \dfrac{2t^2 + 2 - 4t^2 + 2t}{(t^2 + 1)^2}$

$= \dfrac{2 + 2t - 2t^2}{(t^2 + 1)^2} = \dfrac{2(1 + t - t^2)}{(t^2 + 1)^2}.$

TOTAL FOR PAPER: 75 MARKS

GCE Examinations

Test Paper 2

Advanced Level

Core Mathematics C3

Time: 1 hour 30 minutes

Instructions and Information

Candidates may use any calculator allowed by the regulations of their Examination Board.

Full marks are awarded for correct answers to ALL questions.

This paper has eight questions.

You can start working with any question and you must label clearly all parts.

1. $f: x \mapsto -x^2 + 4x - 3$ for $-3 \leq x \leq 3$, $x \in \mathbb{R}$.

   (a) Sketch the graph. (4)

   (b) Indicate the domain and range. (2)

   (c) Draw a mapping diagram. (4)

   (d) Distinguish between domain and co-domain. (2)

2. Factorise $a^3 + b^3$ by finding the quotient $\dfrac{a^3 + b^3}{a + b}$. (4)

   Hence factorise $8x^3 + 27y^3$. (2)

3. The graph of $f(x)$ is shown in Fig. 1.
   Sketch on separate diagrams the graphs:

   (i) $|f(x)|$ (3)

   (ii) $f|(x)|$ (3)

   (iii) $|f|x||$. (3)

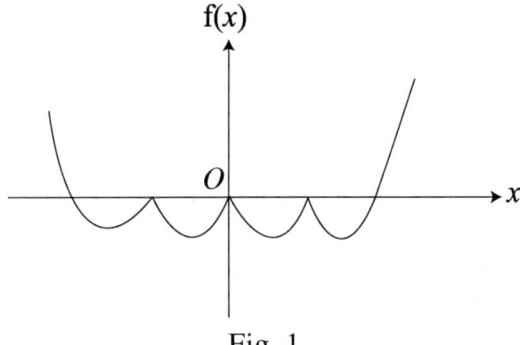

Fig. 1

4. (a) If $f: x \mapsto \sin 2x$ $-2\pi \leq x \leq 2\pi$.
   Sketch the graph. (2)

   (b) (i) Sketch $|\sin 2x|$ $0 \leq x \leq \pi$. (2)

   (ii) Sketch $\sin 2|x|$ $-\pi \leq x \leq \pi$. (2)

   (c) $g: x \mapsto -\dfrac{1}{x}$. Sketch the graph. (1)

   Sketch $|g(x)|$ and $g|x|$. (4)

5. Find without using tables or calculators the value of x for the following:
   (a) $\tan^{-1} 2 + \tan^{-1} 1 = \sin^{-1} x$ (5)
   (b) $\tan^{-1} \frac{1}{2} + \tan^{-1} \frac{1}{3} = \cos^{-1} x$ (6)
   (c) $\tan^{-1} 5 + \tan^{-1} 1 = \tan^{-1} x$. (5)

---

6. Show that $\dfrac{d}{dr}\left[\sin^{-1}\left(\dfrac{a}{r}\right)\right] = -\dfrac{a}{r\sqrt{r^2 - a^2}}.$ (6)

---

7. If $g(x) = 3 \ln x - 1 \quad x > 0$.
   (a) Find $g^{-1}(x)$. (3)
   (b) Sketch $g(x)$ and $g^{-1}(x)$ and find the coordinates of the points of intersections with the axes. (6)

---

8. Sketch the graph $y = \tan^{-1} x \quad -\frac{\pi}{2} \leq x \leq \frac{\pi}{2}$. Sketch the line graph $\frac{x}{5} + \frac{y}{6} = 1$. Show that the x-coordinate of P satisfies the equation

$$x = 5 - \frac{5}{6} \tan^{-1} x.$$ (5)

The iterative formula

$$x_{n+1} = 5 - \frac{5}{6} \tan^{-1} x_n$$

gives the approximation to the solution of this equation.

If $x_0 = 3.8$, find x_1 and x_2 to 5 decimal places. (5)

TOTAL FOR PAPER: 75 MARKS

GCE Examinations

Test Paper 2 Solutions

Advanced Level

Core Mathematics C3

1. (a) f(x) is the image of x. For the domain of the function $-3 \leq x \leq 3$, the elements or members of the domain are $-3, -2, -1, 0, 1, 2, 3$ and the image set of the function can be found by substituting the values of x in

$$f(x) = -x^2 + 4x - 2$$
$$f(-3) = -9 - 12 - 2 = -23$$
$$f(-2) = -4 - 8 - 2 = -14$$
$$f(-1) = -1 - 4 - 2 = -7$$
$$f(0) = -2$$
$$f(1) = -1 + 4 - 2 = 1$$
$$f(2) = -4 + 8 - 2 = 2$$
$$f(3) = -9 + 12 - 2 = 1$$

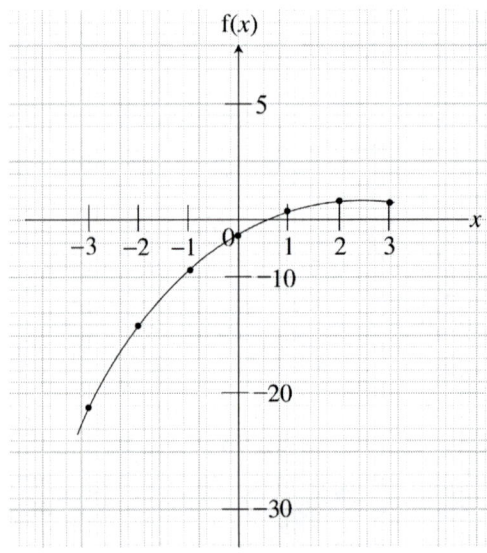

(b) Codomain (Range) (y: $-23 \leq y \leq 2$) $-3 \leq x \leq 3$ domain.

(c) Mapping diagram. Many-to-one-function.

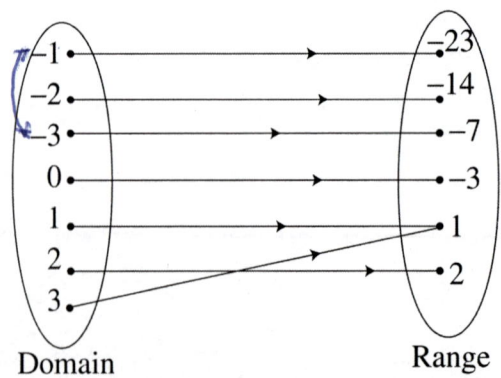

(d) For all real values $-\infty < y \leq 2$ or $(-\infty, 2]$

while the domain: $-\infty < y < \infty$ or $(-\infty, \infty)$.

---

2. $a^3 + b^3$ if $a = -b$ then $-b^3 + b^3 = 0$

$\therefore a + b$ is a factor of $a^3 + b^3$.

$$
\begin{array}{r}
a^2 - ab + b^2 \\
a+b\overline{\smash{\big)}\,a^3 + b^3} \\
\underline{a^3 + a^2b} \\
-a^2b + b^3 \\
\underline{-a^2b - ab^2} \\
ab^2 + b^3 \\
\underline{ab^2 + b^3} \\
0
\end{array}
$$

$\therefore \quad a^3 + b^3 = (a+b)(a^2 - ab + b^2)$

$8x^3 + 27y^3 = (2x + 3y)(4x^2 - 2x3y + 9y^2)$

$\qquad = (2x + 3y)(4x^2 - 6xy + 9y^2).$

3.

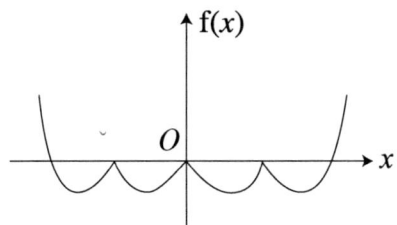

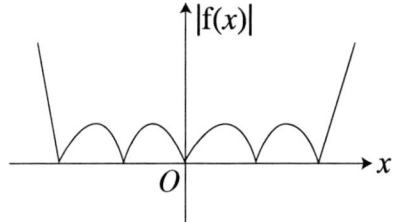

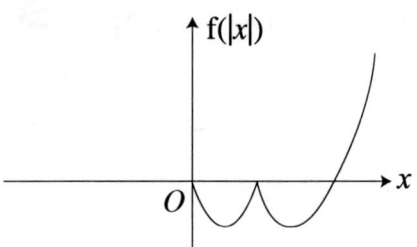

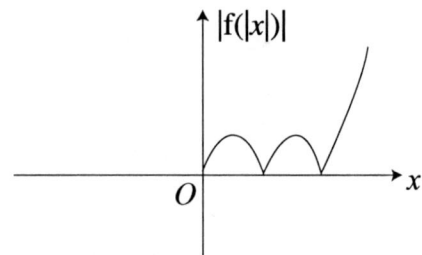

4. (a)

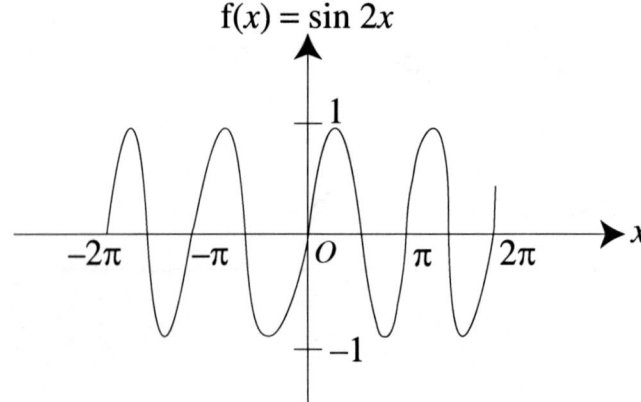

$$f(x) = \sin 2x$$

(b) (i)

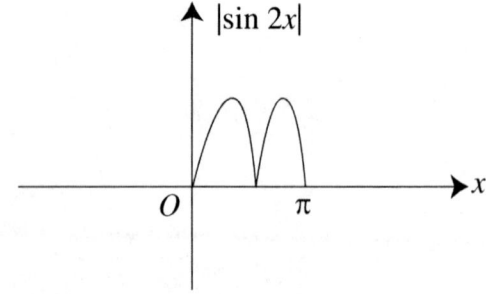

$|\sin 2x|$

(ii)

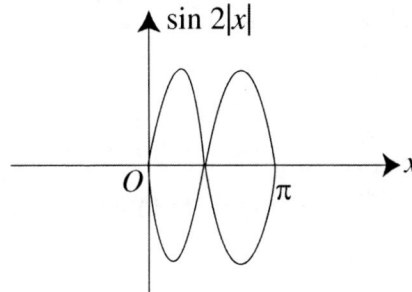

(c)

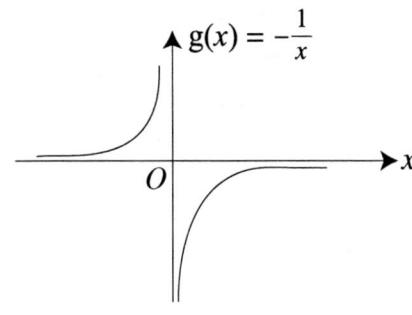

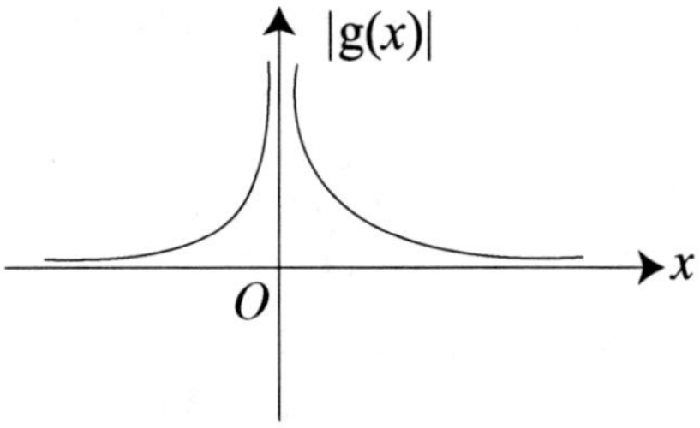

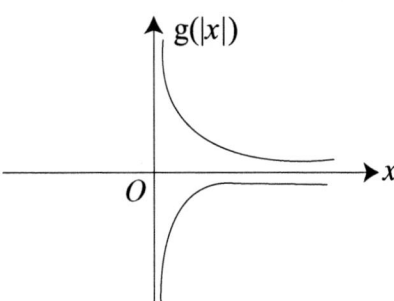

5. (i) $\tan^{-1} 2 + \tan^{-1} 1 = \sin^{-1} x$

$$\theta - \phi = \psi$$

taking tangents on both sides

$$\tan(\theta - \phi) = \tan \psi$$

$$\frac{\tan \theta - \tan \phi}{1 + \tan \theta \tan \phi} = \frac{2 - 1}{1 + 2 \times 1} = \frac{1}{3}$$

$$\tan(\sin^{-1} x) = \frac{1}{3}$$

$$\tan \psi = \frac{1}{3}$$

$$\sin \psi = \boxed{x = \frac{1}{\sqrt{10}}}$$

(ii) $\tan^{-1} \frac{1}{2} + \tan^{-1} \frac{1}{3} = \cos^{-1} x$

$$\tan\left(\tan^{-1} \frac{1}{2} + \tan \tan^{-1} \frac{1}{3}\right) = \tan \cos^{-1} x$$

$$\frac{\tan\left(\tan^{-1} \frac{1}{2}\right) + \tan\left(\tan^{-1} \frac{1}{3}\right)}{1 - \tan\left(\tan^{-1} \frac{1}{2}\right) \tan\left(\tan^{-1} \frac{1}{3}\right)} = \tan(\cos^{-1} x)$$

$$\frac{\frac{1}{2} + \frac{1}{3}}{1 - \frac{1}{2} \times \frac{1}{3}} = \frac{\frac{5}{6}}{\frac{5}{6}} = 1 = \tan(\cos^{-1} x).$$

let $\cos^{-1} x = \theta$

$$\tan \theta = 1$$

$$\tan(\cos^{-1} x) = 1$$

$$\cos^{-1} x = \frac{\pi}{4}$$

$$\cos \frac{\pi}{4} = \frac{1}{\sqrt{2}}$$

$$\therefore \boxed{x = \frac{1}{\sqrt{2}}}$$

(iii)
$$\tan^{-1}5 + \tan^{-1}1 = \tan^{-1}x$$
$$\tan(\tan^{-1}5 + \tan^{-1}1) = \tan(\tan^{-1}x)$$
$$\frac{\tan(\tan^{-1}5) + \tan(\tan^{-1}1)}{1 - \tan(\tan^{-1}5)\tan(\tan^{-1}1)} = x$$
$$\frac{5+1}{1-5\times 1} = x = \frac{6}{-4} = -1.5$$
$$\boxed{x = -1.5}$$

6.
$$y = \sin^{-1}\left(\frac{a}{r}\right)$$
$$\frac{a}{r} = \sin y$$
$$r = \frac{a}{\sin y} = a\operatorname{cosec} y$$
$$\frac{dr}{dy} = -a\operatorname{cosec} y \cot y$$
$$\frac{d}{dr}\left[\sin^{-1}\left(\frac{a}{r}\right)\right] = -\frac{1}{a\operatorname{cosec} y \cot y}$$
$$= -\frac{\sin y \tan y}{a}$$
$$= -\frac{a}{r}\cdot\frac{1}{\sqrt{r^2-a^2}}$$
$$= -\frac{a}{r\sqrt{r^2-a^2}}.$$

---

7. (a)
$$g(x) = 3\ln x - 1 \quad x > 0$$
$$y = 3\ln x - 1$$
$$x = 3\ln y - 1$$
$$3\ln y = x + 1$$
$$\ln y = \frac{1}{3}(x+1)$$
$$y = e^{\frac{1}{3}(x+1)}$$
$$g^{-1}(x) = e^{\frac{1}{3}(x+1)}.$$

(b)

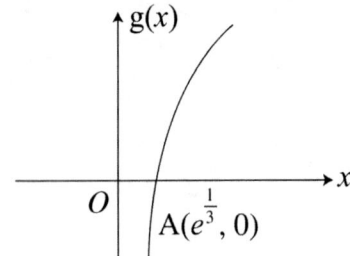

If $g(x) = 0 \Rightarrow 3\ln x - 1 = 0$
$\ln x^3 = 1$
$e^1 = x^3$
$x = e^{\frac{1}{3}} = 1.4.$

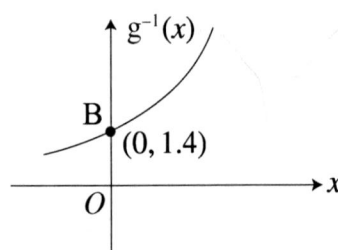

$g^{-1}(x) = e^{\frac{1}{3}(x+1)}$
$g^{-1}(0) = e^{\frac{1}{3}} = 1.4$ to 1 d.p.

8.

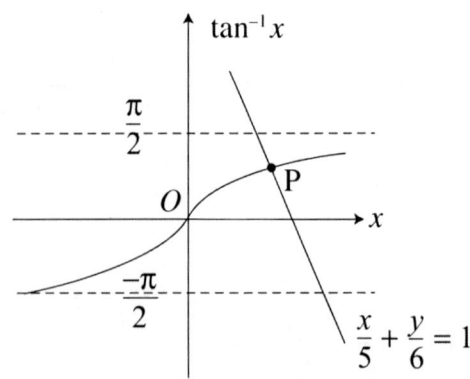

$y = \tan^{-1} x = 6\left(1 - \dfrac{x}{5}\right) = 6 - \dfrac{6x}{5}$

$5\tan^{-1} x + 6x - 30 = 0$

$6x - 30 = -5\tan^{-1} x$

$x - 5 = -\dfrac{5}{6}\tan^{-1} x$

$$\boxed{x_{n+1} = 5 - \dfrac{5}{6}\tan^{-1} x_n}$$

$$x_0 = 3.8$$
$$x_1 = 5 - \tfrac{5}{6}\tan^{-1} 3.8$$
$$= 3.90543949 = 3.90543 \text{ to 5 d.p.}$$
$$x_2 = 5 - \frac{5}{6}\tan^{-1} 3.90543949$$
$$= 3.899892695 = 3.89989 \text{ to 5 d.p.}$$

TOTAL FOR PAPER: 75 MARKS

GCE Examinations

Test Paper 3

Advanced Level

Core Mathematics C3

Time: 1 hour 30 minutes

Instructions and Information

Candidates may use any calculator allowed by the regulations of their Examination Board.

Full marks are awarded for correct answers to ALL questions.

This paper has eight questions.

You can start working with any question and you must label clearly all parts.

1. (a) Determine the inverse function for $f(x) = \frac{3x+1}{2x-3}$ $x \neq \frac{3}{2}$. (4)

   (b) Determine the inverse function for $g(x) = \ln x$ $x > 0$. (2)

   (c) Sketch $g(x)$ and $g'(x)$ on the same graph, indicating the coordinates of the points of intersections with the axes. (4)

---

2. Find the quotient $\dfrac{a^3 - b^3}{a - b}$. (4)

   Hence factorise $8x^3 - 27y^3$. (2)

---

3. (a) A half-wave rectifier voltage waveform is shown in Fig. 1.

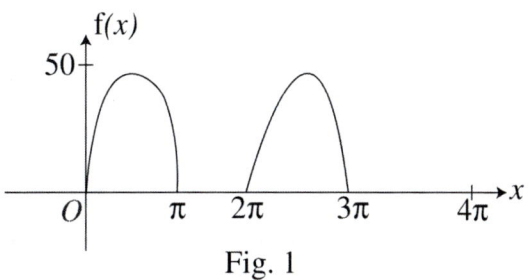

Fig. 1

   Sketch the inverse of $f(x)$. (3)

   (b) A full-wave rectifier voltage waveform is shown in Fig. 2.

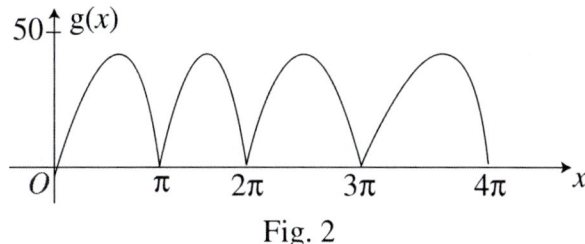

Fig. 2

   Sketch the inverse of $g(x)$. (3)

---

4. Prove that $\sin x = 2 \sin \frac{x}{2} \cos \frac{x}{2}$ and using $\sin^2 \frac{x}{2} + \cos^2 \frac{x}{2} = 1$ show that

$$\sin x = \frac{2 \tan \frac{x}{2}}{1 + \tan^2 \frac{x}{2}}.$$

(5)

and by considering the right angle triangle in Fig. 3.

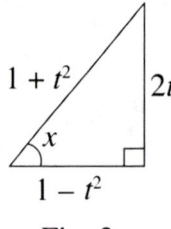

Fig. 3

where $t = \tan \frac{x}{2}$.

Write down the expressions for

(i) $\cos x$ (1)

(ii) $\tan x$ (1)

(iii) $\operatorname{cosec} x$ (1)

(iv) $\sec x$ (1)

(v) $\cot x$. (1)

---

5. Using the graphs of the quadratic function

$$f(x) = ax^2 + bx + c$$

in Fig. 4 and Fig. 5.

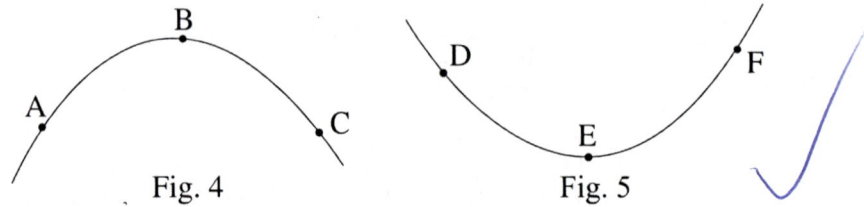

Fig. 4   Fig. 5

Explain the *sign* of the gradients at the points A, B, C, D, E and F.

(i) $\frac{dy}{dx} = f'(x)$

(ii) $\frac{d}{dx}\left(\frac{dy}{dx}\right) = \frac{d^2y}{dx^2} = f''(x)$.

Draw your conclusions. (8)

6. A square plate of 1 m side is used to make a container to hold maximum volume as shown in Fig. 6 by cutting four small squares and turning up the sides.

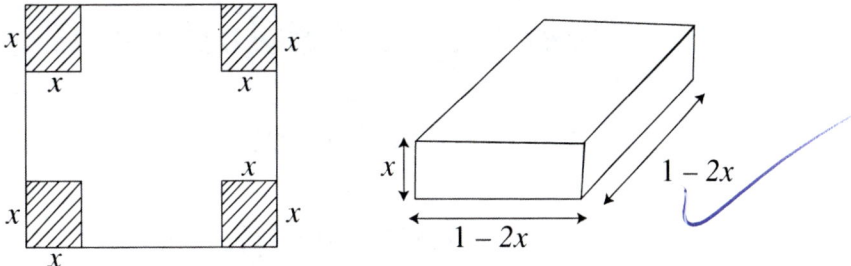

Fig. 6

(a) Determine the value of x for maximum volume. (6)

(b) Calculate the maximum volume justifying that it is maximum. (7)

---

7. Find the first derivatives of the following functions:

(i) $y = \tan x$ (3)

(ii) $y = \cot x$ (3)

(iii) $y = \sec x$ (3)

(iv) $y = \operatorname{cosec} x$. (3)

The first derivatives of $y = \sin x$ and $y = \cos x$ are given $\frac{d}{dx}(\sin x) = \cos x$ and $\frac{d(\cos x)}{dx} = -\sin x$.

---

8. Solve the simultaneous equations

$$y = 2e^x \qquad y + x = 4.$$

Show that the approximate value of the x-coordinate of the point of intersection satisfies the equation

$$x = \ln\left|2 - \frac{x}{2}\right|.$$ (4)

The iterative formula $x_{n+1} = \ln\left|2 - \frac{x_n}{2}\right|$ gives the approximation to the solution of this equation.

If $x_0 = 3$, find $x_1, x_2, x_3, x_4, x_5, x_6$ and x_7 and hence write down the answer for x to 3 decimal places. (6)

TOTAL FOR PAPER: 75 MARKS

GCE Examinations

Test Paper 3 Solutions

Advanced Level

Core Mathematics C3

1. (a) $y = \dfrac{3x+1}{2x-3}$

   replace x for y and y for x and solve for y.
   $$x = \dfrac{3y+1}{2y-3}$$
   $$2yx - 3x = 3y + 1$$
   $$2yx - 3y = 3x + 1$$
   $$y(2x - 3) = 3x + 1$$
   $$y = \dfrac{3x+1}{2x-3}$$
   $$f^{-1}(x) = \dfrac{3x+1}{2x-3} \quad x \neq \dfrac{3}{2}.$$

   (b) $y = \ln x$
   $x = \ln y$
   $y = e^x$
   $\therefore g^{-1}(x) = e^x$
   $g^{-1} \; x \mapsto e^x.$

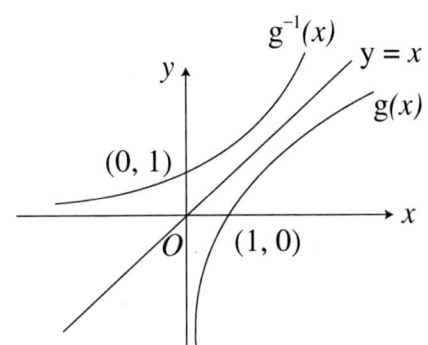

2. $\dfrac{a^3 - b^3}{a - b}$

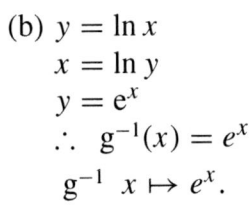

$\therefore a^3 - b^3 = (a - b)(a^2 + ab + b^2)$

$\dfrac{a^3 - b^3}{a - b} = a^2 + ab + b^2$

$8x^3 - 27y^3 = (2x)^3 - (3y)^3 = (2x - 3y)(4x^2 + 6xy + 9y^2).$

3.

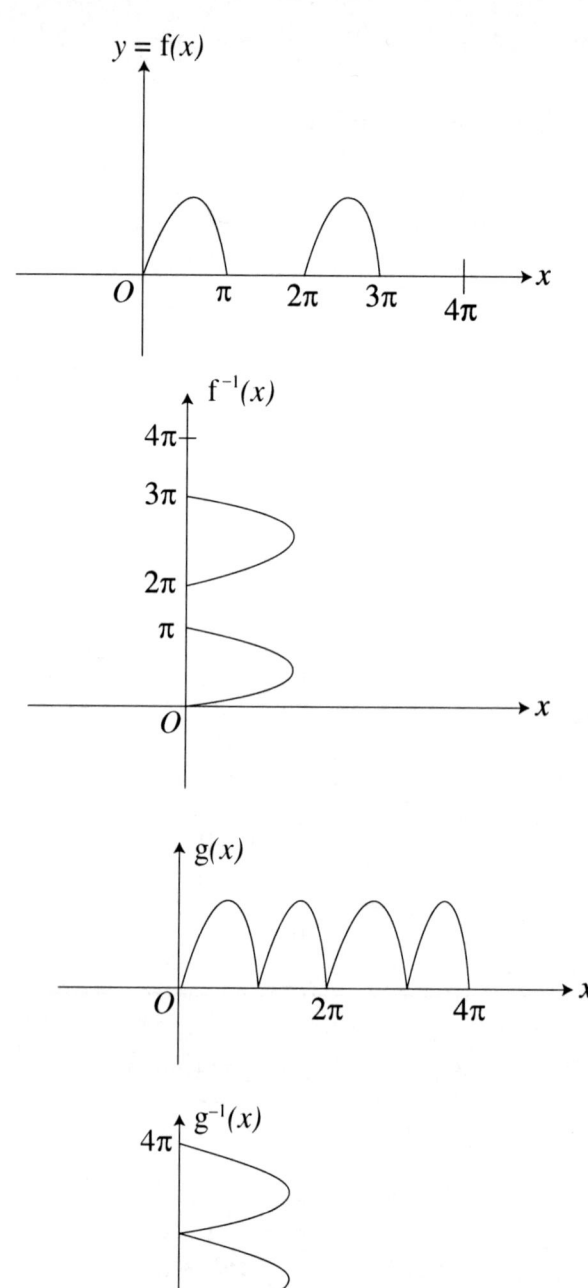

4. (a) $\sin(A+B) = \sin A \cos B + \cos A \sin B$ if $A = B = \dfrac{x}{2}$ then ← GIVEN

$$\sin x = \frac{2 \sin \dfrac{x}{2} \cos \dfrac{x}{2}}{1}$$

$$= \frac{2 \sin \dfrac{x}{2} \cos \dfrac{x}{2}}{\sin^2 \dfrac{x}{2} + \cos^2 \dfrac{x}{2}}$$

$$= \frac{\dfrac{2 \sin \dfrac{x}{2} \cos \dfrac{x}{2}}{\cos^2 \dfrac{x}{2}}}{\dfrac{\sin^2 \dfrac{x}{2} + \cos^2 \dfrac{x}{2}}{\cos^2 \dfrac{x}{2}}}$$

$$= \frac{2 \dfrac{\sin \dfrac{x}{2}}{\cos \dfrac{x}{2}}}{\dfrac{\sin^2 \dfrac{x}{2}}{\cos^2 \dfrac{x}{2}} + 1}$$

$$= \frac{2 \tan \dfrac{x}{2}}{1 + \tan^2 \dfrac{x}{2}} = \frac{2t}{1+t^2}$$

$$\sin x = \frac{2t}{1+t^2}$$

(i) $\cos x = \dfrac{1-t^2}{1+t^2}$

(ii) $\tan x = \dfrac{2t}{1-t^2}$

(iii) $\operatorname{cosec} x = \dfrac{1+t^2}{2t}$

(iv) $\sec x = \dfrac{1+t^2}{1-t^2}$

(v) $\cot x = \dfrac{1-t^2}{2t}$.

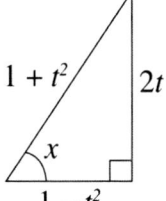

5. (i) The gradients at A, B and C for Fig. 4 are positive, zero and negative, that is,
$$\frac{dy}{dx} > 0, \quad \frac{dy}{dx} = 0, \quad \frac{dy}{dx} < 0.$$

The gradients at D, E and F for Fig. 5 are negative, zero and positive, that is,
$$\frac{dy}{dx} < 0, \quad \frac{dy}{dx} = 0, \quad \frac{dy}{dx} > 0.$$

(ii) $\frac{d}{dx}\left(\frac{dy}{dx}\right) = \frac{d^2y}{dx^2} = f''(x)$ denotes the change of the gradient, for

Fig 4. $\frac{d}{dx}\left(\frac{dy}{dx}\right) < 0$ and for Fig. 5 $\frac{d}{dx}\left(\frac{dy}{dx}\right) > 0$.

For Fig. 4 and Fig. 5 $\frac{dy}{dx} = 0$ for maximum and minimum respectively

$\frac{d^2y}{dx^2} < 0$ for Fig. 4.

$\frac{d^2y}{dx^2} > 0$ for Fig. 5.

6.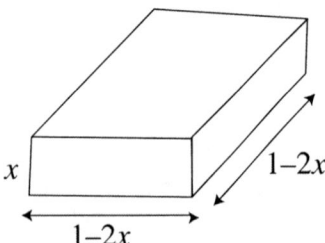

$V = (1 - 2x)(1 - 2x)x$
$\quad = (1 - 4x + 4x^2)x$
$\quad = x - 4x^2 + 4x^3$

$\frac{dV}{dx} = 1 - 8x + 12x^2 = 0$

for turning points

$12x^2 - 8x + 1 = 0$

$x = \frac{8 \pm \sqrt{64 - 48}}{24} = \frac{8 \pm 4\sqrt{4 - 3}}{24}$

$x = \frac{2 \pm 1}{6}$

$x = \frac{1}{2}$ or $x = \frac{1}{6}$.

$$\frac{d^2V}{dx^2} = 24x - 8 \quad f''(x) = 24x - 8$$

$$f''\left(\frac{1}{2}\right) = 12 - 8 = 4 > 0 \quad \text{minimum}$$

$$f''\left(\frac{1}{6}\right) = -4 < 0 \quad \text{maximum}$$

$$V_{\max} = \left(1 - 2 \times \frac{1}{6}\right)^2 \frac{1}{6} = \left(\frac{2}{3}\right)^2 \times \frac{1}{6} = \frac{4}{9} \times \frac{1}{6} = \frac{2}{27}$$

$$= \frac{2}{27} \text{m}^3.$$

---

7. (i) $y = \tan x = \dfrac{\sin x}{\cos x}$

if $y = \dfrac{u}{v}$

the quotient rule $\dfrac{dy}{dx} = \dfrac{\dfrac{du}{dx} v - u \dfrac{dv}{dx}}{v^2}$

$$\frac{dy}{dx} = \frac{\cos x \cos x - \sin x(-\sin x)}{\cos^2 x}$$

$$= \frac{\cos^2 x + \sin^2 x}{\cos^2 x} = \frac{1}{\cos^2 x} = \sec^2 x$$

$$\therefore \boxed{\frac{d}{dx}(\tan x) = \sec^2 x}$$

(ii) $y = \cot x = \dfrac{\cos x}{\sin x}$

$$\frac{dy}{dx} = \frac{-\sin x \sin x - \cos x(\cos x)}{\cos^2 x} = -\frac{\sin^2 x + \cos^2 x}{\sin^2 x}$$

$$= -\frac{1}{\sin^2 x} = -\csc^2 x$$

$$\therefore \boxed{\frac{d}{dx}(\cot x) = -\csc^2 x}$$

(iii) $y = \sec x = \dfrac{1}{\cos x}$

$$\dfrac{dy}{dx} = \dfrac{0 \times \cos x + 1(\sin x)}{\cos^2 x} = \dfrac{\sin x}{\cos^2 x} = \dfrac{\sin x}{\cos x} \cdot \dfrac{1}{\cos x}$$

$$= \tan x \sec x$$

$$\therefore \boxed{\dfrac{d}{dx}(\sec x) = \tan x \sec x}$$

(iv) $y = \operatorname{cosec} x$

$$y = \dfrac{1}{\sin x} \qquad \dfrac{dy}{dx} = \dfrac{0(\sin x) - 1\cos x}{\sin^2 x} = -\dfrac{\cos x}{\sin^2 x}$$

$$= -\cot x \cdot \operatorname{cosec} x \qquad \therefore \boxed{\dfrac{d}{dx}(\operatorname{cosec} x) = -\cot x \operatorname{cosec} x}$$

---

8. $y = 2e^x = -x + 4$

$$e^x = -\dfrac{x}{2} + 2$$

$$x = \ln\left|2 - \dfrac{x}{2}\right|$$

$$x_{n+1} = \ln\left|2 - \dfrac{x_n}{2}\right|$$

$x_0 = 3 \quad x_1 = \ln\left(2 - \dfrac{x_0}{2}\right)$

$x_1 = \ln(0.5) = -0.69314718$

$x_2 = \ln\left(2 - \dfrac{x_1}{2}\right)$

$x_2 = \ln 2.34657359 = 0.852956217$

$x_3 = \ln\left(2 - \dfrac{x_2}{2}\right) = \ln 1.573521891 = 0.45331635$

$x_4 = \ln\left(2 - \dfrac{x_3}{2}\right) = 0.572865803$

$x_5 = 0.53857722 \qquad x_6 = 0.548532532$

$x_7 = 0.545652313 \qquad x_8 = 0.546486456$

$\therefore \quad x = 0.546.$

TOTAL FOR PAPER: 75 MARKS

GCE Examinations

Test Paper 4

Advanced Level

Core Mathematics C3

Time: 1 hour 30 minutes

Instructions and Information

Candidates may use any calculator allowed by the regulations of their Examination Board.

Full marks are awarded for correct answers to ALL questions.

This paper has eight questions.

You can start working with any question and you must label clearly all parts.

1. The growth of current, i, in an inductor is given by the equation

$$i = I(1 - e^{-\frac{t}{\tau}}) \times 10^{-3} \text{A}$$

where I and τ are constants. Determine the rate of growth of the current by differentiating i with respect to t. (2)

Calculate the rates $\dfrac{di}{dt}$ at the times

(i) $t = 0$ (ii) $t = 1 \times 10^{-3}$ s and (iii) 10×10^{-3} s, given that $I = 1A$ and $\tau = 5 \times 10^{-3}$ s. (3)

---

2. Express

$$\frac{1}{(x+1)^3} - \frac{1}{(x+1)^2} + \frac{1}{x+1} + \frac{1}{2x+1} \text{ into a single fraction.}$$ (6)

---

3. If $f(x) = 3e^{\frac{x}{2}} + 2$ and $g(x) = 3\ln x - 1$.

Find

(i) $fg(x)$, hence $fg(1)$. (3)

(ii) $gf(x)$, hence $gf(0)$. (3)

(iii) $(fg)^{-1}(x)$, hence $(fg)^{-1}(1)$. (4)

(iv) $(gf)^{-1}(x)$, hence $(gf)^{-1}(2)$. (4)

---

4. (a) Consider the unit radius circle in Fig. 1

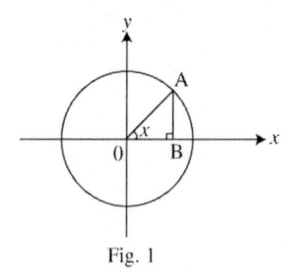

Fig. 1

Write down the definitions for $\sin x$ and $\cos x$ and hence prove the fundamental trigonometric identity $\sin^2 x + \cos^2 x = 1$. (2)

(b) Prove the identities

   (i) $1 + \tan^2 x = \sec^2 x$ (ii) $1 + \cot^2 x = \csc^2 x$

and hence demonstrate these identities on the corresponding unit radius circles of Fig.2 and Fig.3.

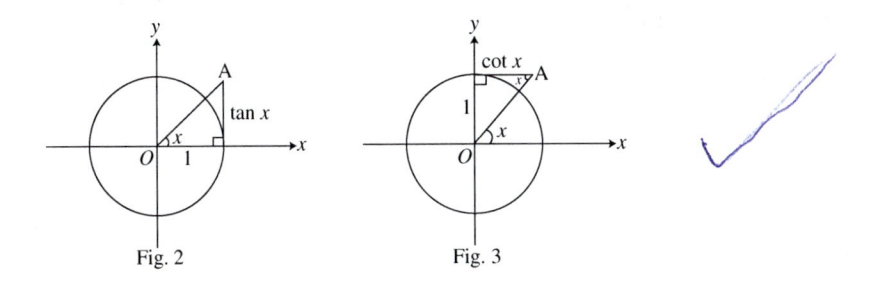

Fig. 2

Fig. 3

Explain what do the lines OA represent in Fig.2 and Fig.3.

(c) Sketch the graphs for $-2\pi \leq x \leq 2\pi$
of (i) $\csc x$ (ii) $\sec x$ (iii) $\cot x$ by labelling clearly the axes. **(6)**

---

5. Fig. 4 shows a circular cylinder to be cut from a solid sphere of radius R as shown in Fig. 4. Show that V, the volume of the cylinder is given by
$$V = 2\pi x^2 (R^2 - x^2)^{\frac{1}{2}}$$ **(4)**

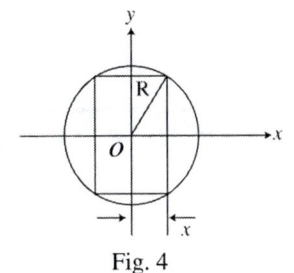

where x is the radius of the base of the cylinder.

Find the maximum volume of the cylinder if x varies, and assume that $\frac{d^2 V}{dx^2}$ is negative. **(6)**

Fig. 4

---

6. (a) Show that
$$\sin^{-1} x + \cos^{-1} x = \frac{\pi}{2}.$$ **(4)**

(b) Show that
$$\tan^{-1}(x+1) + \tan^{-1}(1-x) = \tan^{-1} \frac{2}{x^2}.$$ **(4)**

7. (a) Given that $e^x = 1 + \frac{x}{1!} + \frac{x^2}{2!} + \frac{x^3}{3!} + \ldots + \frac{x^r}{r!} + \ldots$

   where $3! = 1 \times 2 \times 3$.

   Show that $\frac{d}{dx}(e^x) = e^x$. (3)

   Hence find the derivative of e^{kx}.

   (b) if $y = e^x \sin x$,

   find $\frac{dy}{dx}$. (3)

   (c) if $y = 3e^{3x} \sin 2x$,

   find $\frac{dy}{dx}$. (3)

---

8. Sketch the graphs $y = e^{-x}$ and $y = 2x$ for values of $0 < x < 4$, the approximate values of the coordinates of the point of intersection are given $P(0.35, 0.70)$.

   Evaluate the value of $f(x) = e^{-x} - 2x$ when $x = 0.35$ and $x = 0.36$. (4)
   The iterative formula

   $$x_{n+1} = \frac{1}{2}e^{-x_n}$$

   gives the approximation to the solution of this equation.
   If $x_0 = 0.35$, find x_1, x_2 and x_3. (5)

TOTAL FOR PAPER: 75 MARKS

GCE Examinations

Test Paper 4 Solutions

Advanced Level

Core Mathematics C3

1. $i = I(1 - e^{-\frac{t}{\tau}})10^{-3} = I - Ie^{-\frac{t}{\tau}}10^{-3}$

$$\frac{di}{dt} = -I(-\frac{1}{\tau})e^{-\frac{t}{\tau}}10^{-3} = \frac{I}{\tau}e^{-\frac{t}{\tau}}10^{-3}$$

(i) $\dfrac{di}{dt} = \dfrac{1 \times 10^{-3}}{5 \times 10^{-3}}e^{-0} = 0.2$ A/s

(ii) $\dfrac{di}{dt} = \dfrac{1 \times 10^{-3}}{5 \times 10^{-3}}e^{-\frac{1 \times 10^{-3}}{5 \times 10^{-3}}} = 164$ mA/s to 3 s.f.

(iii) $\dfrac{di}{dt} = \dfrac{1 \times 10^{-3}}{5 \times 10^{-3}}e^{-2} = 27.1$ mA/s to 3 s.f.

---

2. $\dfrac{1}{(x+1)^3} - \dfrac{1}{(x+1)^2} + \dfrac{1}{(x+1)} + \dfrac{1}{(2x+1)} = \dfrac{2x+1}{(x+1)^3(2x+1)} - \dfrac{(x+1)(2x+1)}{(x+1)^3(2x+1)}$

$+ \dfrac{(x+1)^2(2x+1)}{(x+1)^3(2x+1)} + \dfrac{(x+1)^3}{(x+1)^3(2x+1)}$

$= \dfrac{2x+1 - (2x^2 + 2x + x + 1) + (x^2 + 2x + 1)(2x+1) + x^3 + 3x^2 + 3x + 1}{(x+1)^3(2x+1)}$

$= \dfrac{2x + 1 - 2x^2 - 2x - x - 1 + 2x^3 + 4x^2 + 2x + x^2 + 2x + 1 + x^3 + 3x^2 + 3x + 1}{(x+1)^3(2x+1)}$

$= \dfrac{3x^3 + 6x^2 + 6x + 2}{(x+1)^3(2x+1)}.$

---

3. (a) $f(x) = 3e^{\frac{x}{2}} + 2, \quad g(x) = 3\ln x - 1$

(i) $fg(x) = 3e^{\frac{3\ln x - 1}{2}} + 2$

$= 3e^{\frac{3}{2}\ln x}e^{-\frac{1}{2}} + 2$

$= 3e^{\ln x^{\frac{3}{2}}}e^{-\frac{1}{2}} + 2$

$fg(x) = \dfrac{3x^{\frac{3}{2}}}{\sqrt{e}} + 2$

$fg(1) = \dfrac{3}{\sqrt{e}} + 2.$

(ii) $gf(x) = 3\ln(3e^{\frac{x}{2}} + 2) - 1$

$gf(0) = 3\ln(3e^0 + 2) - 1$

$= 3\ln 5 - 1.$

(iii) $y = \dfrac{3}{\sqrt{e}} e^{\ln x^{\frac{3}{2}}} + 2$

$y = \dfrac{3}{\sqrt{e}} x^{\frac{3}{2}} + 2$

$x = \dfrac{3}{\sqrt{e}} y^{\frac{3}{2}} + 2$

$\dfrac{3}{\sqrt{e}} y^{\frac{3}{2}} + 2 = x$

$\dfrac{3}{\sqrt{e}} y^{\frac{3}{2}} = x - 2$

$y^{\frac{3}{2}} = \dfrac{(x-2)\sqrt{e}}{3}$

$y = \left[\dfrac{(x-2)\sqrt{e}}{3}\right]^{\frac{2}{3}}$

$(fg)^{-1}(x) = \left[\dfrac{(x-2)\sqrt{e}}{3}\right]^{\frac{2}{3}}$

$(fg)^{-1}(3) = \left(\dfrac{\sqrt{e}}{3}\right)^{\frac{2}{3}}$

$= (0.549573756)^{\frac{2}{3}}$

$= (0.670940472)$

$= 0.671$ to 3 s.f.

(iv) $gf(x) = 3\ln(3e^{\frac{x}{2}} + 2) - 1$

$y = 3\ln(3e^{\frac{x}{2}} + 2) - 1$

$x = 3\ln(3e^{\frac{y}{2}} + 2) - 1$

$\ln(3e^{\frac{y}{2}} + 2) = \dfrac{x+1}{3}$

$3e^{\frac{y}{2}} + 2 = e^{\frac{x+1}{3}}$

$3e^{\frac{y}{2}} + 2 = e^{\frac{x+1}{3}}$

$3e^{\frac{y}{2}} = e^{\frac{x+1}{3}} - 2$

$\dfrac{y}{2} = \ln\left(\dfrac{e^{\frac{x+1}{3}} - 2}{3}\right)$

$y = 2\ln\left(\dfrac{e^{\frac{x+1}{3}} - 2}{3}\right)$

$(gf)^{-1}(x) = 2\ln\left(\dfrac{e^{\frac{x+1}{3}} - 2}{3}\right)$

$gf^{-1}(2) = 2\ln\left(\dfrac{e^{1} - 2}{3}\right)$

$= \ln \dfrac{(e-2)^2}{9} = -2.86$ to 3 s.f.

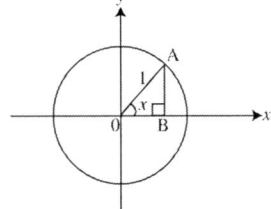

4. (a)

$\sin x = \dfrac{AB}{1} \qquad \cos x = \dfrac{OB}{1}$

∴ AB represents $\sin x$ and OB represents $\cos x$ and using Pythagoras theorem

$OB^2 + AB^2 = OA^2 \Rightarrow \cos^2 x + \sin^2 x = 1$.

(b) $\sin^2 x + \cos^2 x = 1$

dividing each side by $\cos^2 x$ we have

$$\frac{\sin^2 x}{\cos^2 x} + \frac{\cos^2 x}{\cos^2 x} = \frac{1}{\cos^2 x}$$

$$\boxed{\tan^2 + 1 = \sec^2 x}$$

$\sin^2 x + \cos^2 x = 1$

dividing each side by $\sin^2 x$ we have

$$\frac{\sin^2 x}{\sin^2 x} + \frac{\cos^2 x}{\sin^2 x} = \frac{1}{\sin^2 x}$$

$1 + \cot^2 = \csc^2 x$

OA in Fig. 2 represents $\sec x$ and OA in Fig. 3 reprensents $\csc x$ and using Pythagoras Theorem prove the above identities.

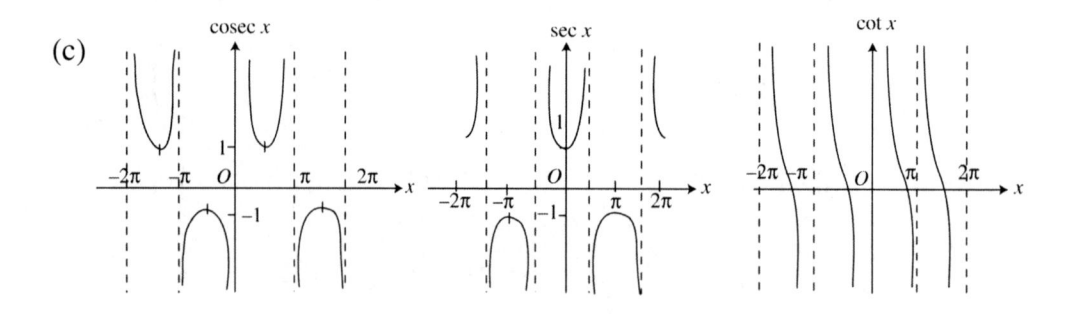

(c)

---

5. $h_1 = \sqrt{R^2 - x^2}$ = half the height of the cylinder

$h = 2\sqrt{R^2 - x^2}$ = height of the cylinder

$V = \pi x^2 h = \pi x^2 2\sqrt{R^2 - x^2}$

$\therefore V = 2\pi x^2 \sqrt{R^2 - x^2}$

$\dfrac{dV}{dx} = 4\pi x \sqrt{R^2 - x^2} + 2\pi x^2 \dfrac{1}{2}(R^2 - x^2)^{\frac{1}{2}}(-2x)$

$\dfrac{dV}{dx} = 0$

$$4\pi x\sqrt{R^2 - x^2} = \frac{2\pi x^3}{\sqrt{R^2 - x^2}}$$

$$2(R^2 - x^2) = x^2$$

$$2(R^2) = 3x^2$$

$$x^2 = \frac{2}{3}R^2$$

$$x = \sqrt{\frac{2}{3}}R$$

$$V_{max} = 2\pi \frac{2}{3}R^2 \sqrt{R^2 - \frac{2}{3}R^2}$$

$$= 2\pi \frac{2}{3}R^2 \frac{1}{\sqrt{3}}R$$

$$= \frac{4\pi}{3\sqrt{3}}R^3.$$

---

6. (a) Proof $\sin^{-1} x + \cos^{-1} x = \frac{\pi}{2}$

let $\sin^{-1} x = \theta \Rightarrow \sin\theta = x$

using the right angle triangle

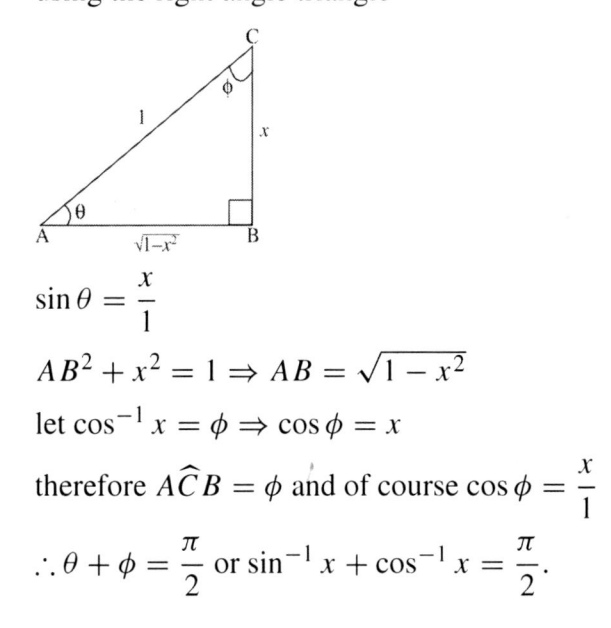

$\sin\theta = \frac{x}{1}$

$AB^2 + x^2 = 1 \Rightarrow AB = \sqrt{1 - x^2}$

let $\cos^{-1} x = \phi \Rightarrow \cos\phi = x$

therefore $A\widehat{C}B = \phi$ and of course $\cos\phi = \frac{x}{1}$

$\therefore \theta + \phi = \frac{\pi}{2}$ or $\sin^{-1} x + \cos^{-1} x = \frac{\pi}{2}.$

b) $\tan^{-1}(1+x) + \tan^{-1}(1-x) = \tan^{-1}\frac{2}{x^2}$.

Taking tangents on both sides

$$\tan\left(\tan^{-1}(1+x) + \tan^{-1}(1-x)\right) = \tan\left(\tan^{-1}\frac{2}{x^2}\right)$$

$$\text{LHS} = \frac{\tan\tan^{-1}(1+x) + \tan\tan^{-1}(1-x)}{1 - \tan\tan^{-1}(1+x)\tan\tan^{-1}(1-x)} = \frac{1+x+1-x}{1-(1+x)(1-x)}$$

$$= \frac{2}{1-1+x^2} = \frac{2}{x^2}$$

$$\text{RHS} = \tan\left(\tan^{-1}\frac{2}{x^2}\right) = \frac{2}{x^2}$$

$\therefore$ LHS = RHS.

---

7. (a) $\dfrac{d}{dx}(e^x) = 0 + \dfrac{1}{1!} + \dfrac{2x}{2!} + \dfrac{3x^2}{3!} + \ldots + \dfrac{rx^{r-1}}{r!} + \ldots$

$$= 1 + \frac{x}{1!} + \frac{x^2}{2!} + \frac{x^3}{3!} + \ldots + \frac{x^r}{r!} + \ldots$$

$$= e^x.$$

$y = e^{kx}$

let $u = kx \qquad \dfrac{du}{dx} = k$

$y = e^u \qquad \dfrac{dy}{du} = e^u$

$\therefore \dfrac{dy}{du} \cdot \dfrac{du}{dx} = \dfrac{dy}{dx} = ke^u = ke^{kx}$

$\therefore \dfrac{d}{dx}(e^{kx}) = ke^{kx}$

(b) $y = uv$ where u and v are functions of x

$$\frac{dy}{dx} = \frac{du}{dx}v + u\frac{dv}{dx}$$

$y = e^x \sin x$

$$\frac{dy}{dx} = e^x \sin x + e^x \cos x = e^x(\sin x + \cos x)$$

$$\frac{d}{dx}(e^x \sin x) = e^x(\sin x + \cos x).$$

(c) $y = 3e^{3x} \sin 2x$

$\dfrac{dy}{dx} = 9e^{3x} \sin 2x + 3e^{3x} 2 \cos 2x = 3e^{3x}(3 \sin 2x + 2 \cos 2x).$

8.

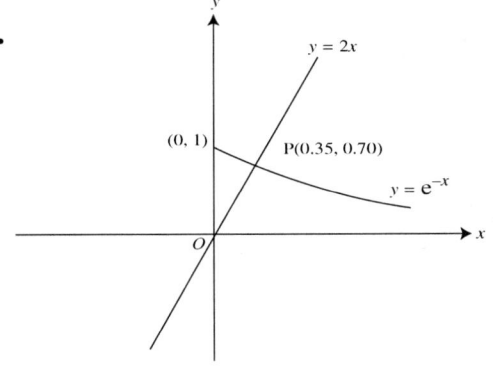

$f(x) = e^{-x} - 2x$

$f(0.35) = e^{-0.35} - 2 \times 0.35 = 0.704688085 - 0.7 = 0.004688085$

$f(0.36) = e^{-0.36} - 2 \times 0.36 = 0.697676326 - 0.72 = -0.022323674$

change of sign

$2x = e^{-x}$

$$\boxed{x_{n+1} = \dfrac{1}{2} e^{-x_n}}$$

$x_1 = \dfrac{1}{2} e^{x_0} = \dfrac{1}{2} e^{-0.35}$

$= \dfrac{1}{2} \times 0.704688085 = 0.352344043$

$x_2 = \dfrac{1}{2} e^{-0.352344043}$

$= 0.351519102$

$x_3 = \dfrac{1}{2} e^{-0.351519102}$

$= 0.351809204.$

TOTAL FOR PAPER: 75 MARKS

GCE Examinations

Test Paper 5

Advanced Level

Core Mathematics C3

Time: 1 hour 30 minutes

Instructions and Information

Candidates may use any calculator allowed by the regulations of their Examination Board.

Full marks are awarded for correct answers to ALL questions.

This paper has eight questions.

You can start working with any question and you must label clearly all parts.

1. f(x) shows a sketch for ≥ 0 in fig. 1

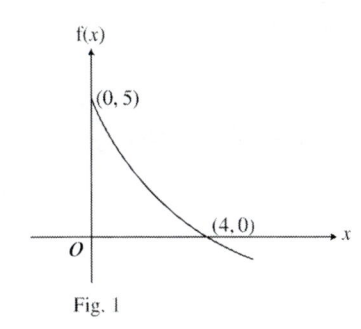

Fig. 1 (graph with points (0, 5) and (4, 0))

The graph intersects the x and y-axes at (4, 0) and (0, 5) respectively.
In separate diagrams sketch the curves with equation

(a) $y = f^{-1}(x)$ (3)

(b) $y = 2f(2x)$. (2)

---

2. if $\dfrac{x^3}{(x+4)(x-5)} \equiv x + A + \dfrac{Bx + C}{(x+4)(x-5)}$

find the integer values of A, B and C. (6)

---

3. The length of the arc s is given by $s = r\theta$. The area of the triangle ABO in Fig. 2 is $\Delta_1 = \frac{1}{2}r^2 \sin\theta$ and that of the sector is AOB $\Delta_2 = \frac{1}{2}r^2\theta$.

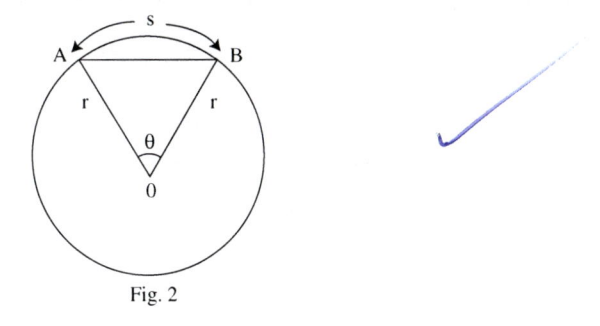

Fig. 2

Express the area of the triangle and sector in terms of s. (2)

Find the maximum area of the triangle Δ_1 assuming r is a constant and s is variable. (4)

Verify that the area is maximum. (4)

What is the area of the sector for this condition. (4)

4. (a) Simplify the following trigonometric expressions:

   (i) $\dfrac{1}{\operatorname{cosec} x + \cot x} + \dfrac{1}{\operatorname{cosec} x - \cot x}$ (2)

   (ii) $\dfrac{1}{\cot x - \operatorname{cosec} x} + \dfrac{1}{\cot x + \operatorname{cosec} x}$ (2)

   (iii) $\dfrac{1}{\sec x + \operatorname{cosec} x} + \dfrac{1}{\operatorname{cosec} x - \sec x}$ (2)

   (b) If $\tan(x+y) = \dfrac{\tan x + \tan y}{1 - \tan x \tan y}$

   show that $\tan(A+B+C) = \dfrac{\tan A + \tan B + \tan C - \tan A \tan B \tan C}{1 - (\tan A \tan B + \tan B \tan C + \tan A \tan C)}$. (5)

5. The quadratic equation
$$x^2 + 4x - 3 = 0$$
is required to be solved by using two iterative methods in order to find the roots of the equation to 2 decimal places. Show that an iterative methods is
$$x_{n+1} = \dfrac{3 - x_n^2}{4}.$$ (5)

Use $x_0 = 1$ in both methods in order to find the solutions. (5)

Check your answers using the quadratic formula. (2)

6. (a) Evaluate the angle $\tan^{-1} 2 + \tan^{-1} 3$ without the use of a calculator. (5)

   (b) Show that $\tan^{-1} + \tan^{-1}\dfrac{1}{2} = \dfrac{\pi}{2}$ (4)

   and in general $\tan^{-1} x + \tan^{-1}\dfrac{1}{x} = \dfrac{\pi}{2}$. (4)

7. (a) if $f(x) = \dfrac{3x+5}{x^2+1}$ determine $f'(x)$. (4)

   (b) if $g(x) = \sqrt{x^2+1}$ determine $g'(x)$. (4)

8. Given that $y = \tan(x + \dfrac{\pi}{4})$, show that $\dfrac{d^2y}{dx^2} = 2y\dfrac{dy}{dx}$. (5)

   Show also that

   $\dfrac{d^3y}{dx^3} = 2\left(\dfrac{dy}{dx}\right)^2 + 4y^2 \dfrac{dy}{dx} = 2\sec^4\left(x + \dfrac{\pi}{4}\right) + 4\tan^2\left(x + \dfrac{\pi}{4}\right)\sec^2\left(x + \dfrac{\pi}{4}\right)$. (5)

TOTAL FOR PAPER: 75 MARKS

GCE Examinations

Test Paper 5 Solutions

Advanced Level

Core Mathematics C3

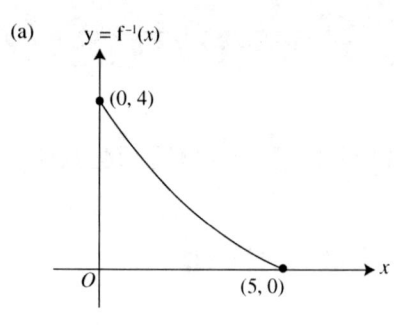

1.

(a) $y = f^{-1}(x)$, curve from $(0, 4)$ to $(5, 0)$.

(b) $y = 2f(2x)$, curve from $(0, 10)$ to $(2, 0)$.

2.

$$\begin{array}{r}
x + 1 \\
x^2 - x - 20 \overline{\smash{\big)} x^3 } \\
\underline{x^3 - x^2 - 20x} \\
x^2 + 20x \\
\underline{x^2 - x - 20} \\
21x + 20
\end{array}$$

$$\therefore \frac{x^3}{(x+4)(x-5)} = x + 1 + \frac{21x + 20}{x^2 - x - 20}$$

$$A = 1, B = 21, C = 20.$$

---

3. $\Delta_1 = \frac{1}{2}r^2 \sin\theta = \frac{1}{2}r^2 \sin(\frac{s}{r})$, $\Delta_2 = \frac{1}{2}r^2(\frac{s}{r}) = \frac{1}{2}rs$. For maximum area $\frac{d\Delta_1}{ds} = 0$

$$\frac{d\Delta_1}{ds} = \frac{1}{r} \cdot \frac{1}{2}r^2 \cos\left(\frac{s}{r}\right) = 0$$

$$\cos\left(\frac{s}{r}\right) = 0 = \cos\frac{\pi}{2}$$

$$\therefore s = r\frac{\pi}{2}$$

$$\frac{d^2\Delta_1}{ds^2} = \frac{1}{2}r^2 \left(\frac{1}{r}\right)\left(\frac{1}{r}\right)\left(-\sin\left(\frac{s}{r}\right)\right)$$

$$= -\frac{1}{2}\sin\left(\frac{s}{r}\right) = -\frac{1}{2}\sin\left(r\frac{\pi}{2}\frac{1}{r}\right)$$

$$= -\frac{1}{2}\sin\frac{\pi}{2} = -\frac{1}{2}$$

which indicates that Δ_1 is maximum

$$\Delta_1 \text{ max} = \frac{1}{2}r^2 \sin\left(\frac{s}{r}\right)$$

$$= \frac{1}{2}r^2 \sin r\frac{\frac{\pi}{2}}{r} = \frac{1}{2}r^2$$

$$\Delta_2 \text{ max} = \frac{1}{2}r^2\theta = \frac{1}{2}r^2\frac{s}{r} = \frac{1}{2}rs = \frac{1}{2}r\cdot\frac{r\pi}{2} = \frac{r^2\pi}{4}$$

$$\boxed{\Delta_2 \text{ max} = \frac{r^2\pi}{4}}$$

---

4. (a) (i) $\dfrac{1}{\operatorname{cosec} x + \cot x} + \dfrac{1}{\operatorname{cosec} x - \cot x}$

$$= \frac{\operatorname{cosec} x - \cot x}{\operatorname{cosec}^2 x - \cot^2 x} + \frac{\operatorname{cosec} x + \cot x}{\operatorname{cosec}^2 x - \cot^2 x}$$

$$= \frac{2\operatorname{cosec} x}{\operatorname{cosec}^2 x - \cot^2 x} = 2\operatorname{cosec} x.$$

(ii) $\dfrac{1}{\cot x - \operatorname{cosec} x} + \dfrac{1}{\cot x + \operatorname{cosec} x}$

$$= \frac{\cot x + \operatorname{cosec} x + \cot x - \operatorname{cosec} x}{\cot^2 x - \operatorname{cosec}^2 x}$$

$$= \frac{2\cot x}{\cot^2 x - \operatorname{cosec}^2 x} = -2\cot x.$$

(iii) $\dfrac{1}{\sec x + \operatorname{cosec} x} + \dfrac{1}{\operatorname{cosec} x - \sec x}$

$= \dfrac{\operatorname{cosec} x - \sec x + \sec x + \operatorname{cosec} x}{\operatorname{cosec}^2 x - \sec^2 x}$

$= \dfrac{2 \operatorname{cosec} x}{\dfrac{1}{\sin^2 x} - \dfrac{1}{\cos^2 x}} = \dfrac{2 \operatorname{cosec} x}{\dfrac{\cos^2 x - \sin^2 x}{\sin^2 x \cos^2 x}} = \dfrac{2 \sin x \cos^2 x}{\cos 2x}$

(b) $\tan(A + B + C) = \tan\left[(A + B) + C\right] = \dfrac{\tan(A+B) + \tan C}{1 - \tan(A+B)\tan C}$

$= \dfrac{\dfrac{\tan A + \tan B}{1 - \tan A \tan B} + \tan C}{1 - \dfrac{\tan A + \tan B}{1 - \tan A \tan B}\tan C}$

$= \dfrac{\tan A + \tan B + \tan C - \tan A \tan B \tan C}{1 - (\tan A \tan B + \tan A \tan C + \tan B \tan C)}.$

5.

Method I	Method II

Method I:

$x^2 + 4x - 3 = 0$

$x^2 = 3 - 4x$

$x = \dfrac{3 - 4x}{x}$

$x_{n+1} = \dfrac{3 - 4x_n}{x_n}$

$x_0 = 1$

$x_1 = \dfrac{3 - 4}{1} = -1$

$x_2 = \dfrac{3 - 4(-1)}{-1} = -7$

$x_3 = \dfrac{3 - 4(-7)}{-7}$

$= \dfrac{31}{-7} = -4.42857...$

Method II:

$x^2 + 4x - 3 = 0$

$4x = 3 - x^2$

$x = \dfrac{3 - x^2}{4}$

$x_{n+1} = \dfrac{3 - x_n^2}{4}$

$x_0 = 1$

$x_1 = \dfrac{1}{2}$

$x_2 = 0.6875$

$x_3 = 0.63183...$

$x_4 = -4.6774...$ $x_4 = 0.65019...$

$x_5 = -4.6413$ $x_5 = 0.6443...$

$x_6 = -4.6463$ $x_6 = 0.6462...$

$x_7 = -46456$ $\boxed{x = 0.65}$

$\boxed{x = -4.65}$

$$x = \frac{-b \pm \sqrt{b^2 - 4ac}}{2a}$$

$$x = \frac{-4 \pm \sqrt{16 + 12}}{2}$$

$$x = \frac{-4 \pm \sqrt{28}}{2}$$

$$x = \frac{-4 - \sqrt{28}}{2} = -4.65 \text{ to 2 d.p.}$$

$$\text{or } x = \frac{-4 + \sqrt{28}}{2} = -0.65 \text{ to 2 d.p.}$$

---

6. (a) Let $\tan^{-1} 2 + \tan^{-1} 3 = \theta + \phi$

where $\tan^{-1} 2 = \theta \Rightarrow \tan\theta = 2$

and $\tan^{-1} 3 = \phi \Rightarrow \tan\phi = 3$.

Taking tangents on both sides

$$\tan(\tan^{-1}2 + \tan^{-1}3) = \frac{\tan\tan^{-1}2 + \tan\tan^{-1}3}{1 - \tan\tan^{-1}2 \tan\tan^{-1}3} = \frac{2+3}{1 - 2 \times 3} = \frac{5}{-5} = -1$$

$\tan(\theta + \phi) = -1 = \tan\dfrac{3\pi}{4}$ the principal value

$\therefore \tan^{-1}2 + \tan^{-1}3 = \dfrac{3\pi}{4}$.

(b) Let $\tan^{-1}2 + \tan^{-1}\dfrac{1}{2} = \phi$

$\tan\left(\tan^{-1}2 + \tan^{-1}\dfrac{1}{2}\right) = \tan\phi$

$$\frac{\tan\tan^{-1}2 + \tan\tan^{-1}\frac{1}{2}}{1 - \tan\tan^{-1}2\tan\tan^{-1}\frac{1}{2}} = \frac{2 + \frac{1}{2}}{1 - 2\times\frac{1}{2}} = \frac{2.5}{0}$$

$$\tan\phi = \infty$$

$$\therefore \phi = \frac{\pi}{2} \qquad \therefore \tan^{-1}2 + \tan^{-1}\frac{1}{2} = \frac{\pi}{2}$$

$$\tan^{-1}x + \tan^{-1}\frac{1}{x} = \theta$$

$$\tan\left(\tan^{-1}x + \tan^{-1}\frac{1}{x}\right) = \tan\theta$$

$$\frac{x + \frac{1}{x}}{1 - x\cdot\frac{1}{x}} = \frac{x + \frac{1}{x}}{0} = \infty \Rightarrow \tan\theta = \infty$$

$$\therefore \theta = \frac{\pi}{2}$$

$$\tan^{-1} + \tan^{-1}\frac{1}{x} = \frac{\pi}{2}.$$

---

7. (a) $\quad f(x) = \dfrac{3x+5}{x^2+1}$

$$f'(x) = \frac{3(x^2+1) - (3x+5)2x}{(x^2+1)^2}$$

$$= \frac{3x^2 + 3 - 6x^2 - 10x}{(x^2+1)^2} = \frac{-3x^2 - 10x + 3}{(x^2+1)^2}.$$

(b) $\quad g(x) = \sqrt{x^2+1} = (x^2+1)^{\frac{1}{2}} = u^{\frac{1}{2}}$

Let $u = x^2 + 1 \Rightarrow \dfrac{du}{dx} = 2x$

$$g'(x) = \frac{1}{2}u^{-\frac{1}{2}}\frac{du}{dx} = \frac{1}{2}u^{-\frac{1}{2}}2x$$

$$= xu^{-\frac{1}{2}} = \frac{x}{\sqrt{u}} = \frac{x}{\sqrt{x^2+1}}$$

$$\therefore g'(x) = \frac{x}{\sqrt{x^2+1}}.$$

8. $y = \tan\left(x + \dfrac{\pi}{4}\right)$

$\dfrac{dy}{dx} = \sec^2\left(x + \dfrac{\pi}{4}\right)$

$\dfrac{d^2y}{dx^2} = 2\sec\left(x + \dfrac{\pi}{4}\right)\sec\left(x + \dfrac{\pi}{4}\right)\tan\left(x + \dfrac{\pi}{4}\right)$

$\qquad = 2\sec^2\left(x + \dfrac{\pi}{4}\right)\tan\left(x + \dfrac{\pi}{4}\right)$

$\qquad = 2\dfrac{dy}{dx}\tan\left(x + \dfrac{\pi}{4}\right)$

$\boxed{\dfrac{d^2y}{dx^2} = 2y\dfrac{dy}{dx}}$

$\dfrac{d^2y}{dx^2} = 2y\dfrac{dy}{dx}$

$\dfrac{d^3y}{dx^3} = 2\dfrac{dy}{dx}\dfrac{dy}{dx} + 2y\dfrac{d^2y}{dx^2}$

$\qquad = 2\left(\dfrac{dy}{dx}\right)^2 + 2y \cdot 2y\dfrac{dy}{dx}$

$\qquad = 2\left(\dfrac{dy}{dx}\right)^2 + 4y^2\dfrac{dy}{dx}$

$\qquad = 2\sec^4\left(x + \dfrac{\pi}{4}\right) + 4\tan^2\left(x + \dfrac{\pi}{4}\right)\sec^2\left(x + \dfrac{\pi}{4}\right).$

TOTAL FOR PAPER: 75 MARKS

GCE Examinations

Test Paper 6

Advanced Level

Core Mathematics C3

Time: 1 hour 30 minutes

Instructions and Information

Candidates may use any calculator allowed by the regulations of their Examination Board.

Full marks are awarded for correct answers to ALL questions.

This paper has eight questions.

You can start working with any question and you must label clearly all parts.

1. The functions h and k are defined by h: $x \mapsto -x + 3$ and k: $x \mapsto 3x - a$ where $x \in \mathbb{R}$ and a is a constant.

   (a) Find a if hk = kh. (2)

   (b) Find the values of

   (i) kk if $a = 1$ (2)

   (ii) hh (2)

   (iii) hkh if $a = 3$ (2)

   (iv) khk if $a = 1$ (2)

   (v) kkk if $a = 2$. (2)

2. (a) if Sketch the exponential functions on the same graph:

   (i) $y = 2^x$ (ii) $y = e^x$ (iii) 3^x (3)

   where $2 < e < 3$.

   (b) Sketch $f(x) = \ln(x + 2)$ and find the coordinates of the points of intersections with the axes. (2)

   (c) Find $f^{-1}(x)$ and determine the coordinates of the point intersecting the axes. (3)

3. Determine the coordinates of the maximum and minimum points for the following functions:

   (a) $f(x) = x^2 + 4x + 8$

   (b) $g(x) = x^2 - 4x + 7$

   (c) $h(x) = -x^2 - 6x - 7$

   (d) $p(x) = -x^2 + 6x - 8$.

   Sketch the graphs and indicate the stationary values on a single graph. (10)

4. An iterative formula for solving simultaneously the functions $f(x) = \ln x$ and $\frac{x}{5} + \frac{y}{3} = 1$ is given by

   $x_{n+1} = 5\left(1 - \frac{1}{3} \ln x_n\right)$.

   If $x_0 = 3$, find x_1, x_2, x_3, x_4 and work out the approximate solution in 2 decimal places. (5)

5. Express $\sqrt{3}\cos x + \sqrt{2}\sin x$ in the form $R\cos(x+\alpha)$. (4)

   Hence sketch the graph indicating the coordinates of the points of intersections with axes for one cycle and the maximum and minimum. (4)

6. Use the compound formula to find the values of the following trigonometric ratios:

   (a) $\sin 15° = \frac{\sqrt{2}}{4}(\sqrt{3}-1)$ (3)

   (b) $\cos 75° = \frac{\sqrt{2}}{4}(\sqrt{3}-1) = \sin 15°$ (3)

   (c) $\tan 15° = 2 - \sqrt{3}$ (3)

   (d) $\cot 75° = \tan 15° = 2 - \sqrt{3}$. (3)

7. (a) Given that $f(x) = x^3\sqrt{2x^3 - 1}$, find $f'(x)$. (4)

   (b) Given that $g(x) = \dfrac{\cos 2x}{x^2 + 1}$, find $g'(x)$. (4)

8.

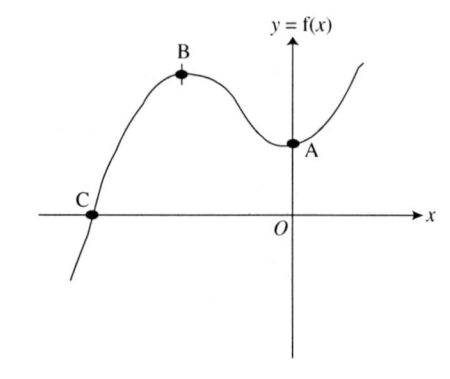

   $A(0, 1)$ $B(-1, 2)$ $C(-3, 0)$

   Sketch separately the graphs of

   (i) $y = f(-x)$ (3)

   (ii) $y = f(x+3)$ (3)

   (iii) $y = -f(x)$ (3)

   (iv) $y = f(x+1)$. (3)

   TOTAL FOR PAPER: 75 MARKS

GCE Examinations

Test Paper 6 Solutions

Advanced Level

Core Mathematics C3

1. (a) hk: $x = -(3x - a) + 3 = -3x + a + 3$

kh: $x = 3(-x + 3) - a = -3x + 9 - a$

$$-3x + a + 3 = -3x + 9 - a$$
$$2a = 6$$
$$\boxed{a = 3}$$

(b) (i) kk: $x = 3(3x - a) - a = 9x - 3a - a$
$$= 9x - 4a$$
$$= 9x - 4.$$

(ii) h: $x \mapsto -x + 3$ \quad hh: $x \mapsto -(-x + 3) + 3 = x - 3 + 3 = x$
$$\therefore \text{hh}: x \mapsto x.$$

(iii) h: $x \mapsto -x + 3$ \quad k: $x \mapsto 3x - 3$

hk: $x \mapsto [-(3x - 3) + 3] = -3x + 3 + 3 = -3x + 6$

hk: $x \mapsto -3x + 6$

hkh: $x \mapsto [-3(-x + 3) + 6] = 3x - 9 + 6 = 3x - 3$

hkh: $x \mapsto 3x - 3.$

(iv) k: $x \mapsto 3x - 1$ \quad h: $x \mapsto -x + 3$

kh: $x \mapsto 3(-x + 3) - 1 = -3x + 9 - 1 = -3x + 8$

kh: $x \mapsto -3x + 8$

khk: $x \mapsto -3(3x - 1) + 8 = -9x + 3 + 8 = -9x + 11$

khk: $x \mapsto -9x + 11.$

(v) k: $x \mapsto 3x - 2$ \quad kk: $x \mapsto 3(3x - 2) - 2 = 9x - 6 - 2 = 9x - 8$

$\therefore$ kkk: $x \mapsto 9(3x - 2) - 8 = 27x - 26.$

2. (a)

(graph showing 2^x, e^x, 3^x passing through $(0, 1)$)

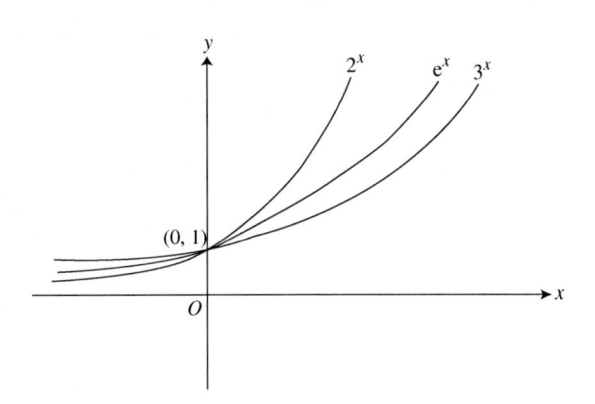

(b)

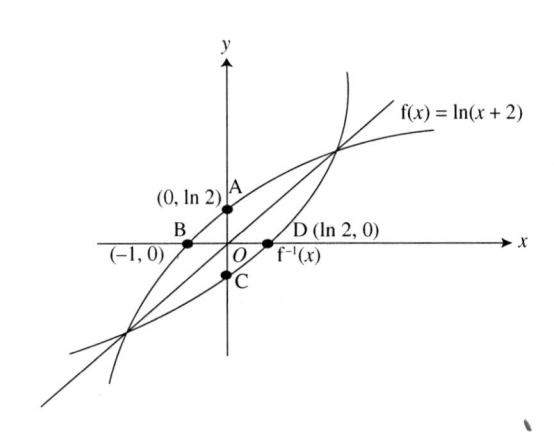

(graph showing $f(x) = \ln(x+2)$ and $f^{-1}(x)$ with points $A(0, \ln 2)$, $B(-1, 0)$, $D(\ln 2, 0)$, C)

$f(0) = \ln 2$

$\ln(x + 2) = 0$

$e^0 = x + 2$

$x = 1 - 2 = -1$

$\boxed{A(0, \ln 2)}$

$\boxed{B(-1, 0)}$

(c) $y = \ln(x + 2)$

replace y with x and x with y and solve for y

$x = \ln(y + 2)$

$\therefore e^x = y + 2$

$y = e^x - 2$

$f^{-1}(x) = e^x - 2$

$C(0, -1)$

$e^x - 2 = 0$

$e^x = 2$

$x = \ln 2$

$\boxed{D(\ln 2, 0)}$

$\boxed{C(0, -1)}$

3. (i) $f(x) = x^2 + 4x + 8$ (ii) $g(x) = x^2 - 4x + 7$
   $f'(x) = 2x + 4$ $g'(x) = 2x - 4$
   $f''(x) = 2$ minimum $g''(x) = 2$ minimum

   (iii) $h(x) = -x^2 - 6x - 7$ (iv) $p(x) = -x^2 + 6x - 8$
   $h'(x) = -2x - 6$ $p'(x) = -2x + 6$
   $h''(x) = -2$ maximum $p''(x) = -2$ maximum.

   For turning points the first derivatives are equal to zero.

   (i) $f'(x) = 2x + 4 = 0 \Rightarrow x = -2$ for minimum

   (ii) $g'(x) = 2x - 4 = 0 \Rightarrow x = 2$ for minimum

   (iii) $h'(x) = -2x - 6 = 0 \Rightarrow x = -3$ for maximum

   (iv) $p'(x) = -2x + 6 = 0 \Rightarrow x = 3$ for maximum

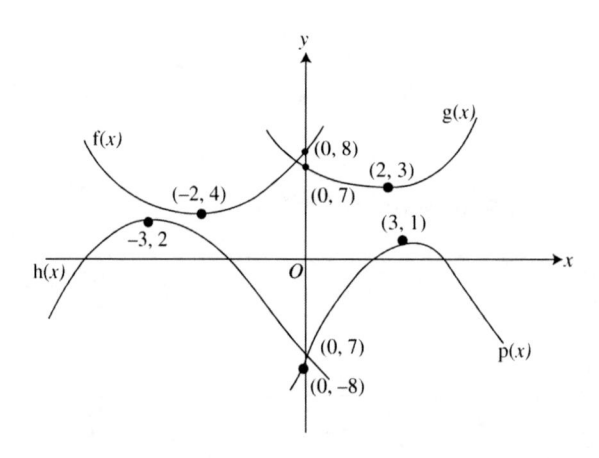

   $f(-2) = 4 - 8 + 8 = 4$

   $g(2) = 4 - 8 + 7 = 3$

   $h(-3) = -9 + 18 - 7 = 2$

   $p(3) = -9 + 18 - 8 = 1$.

4.

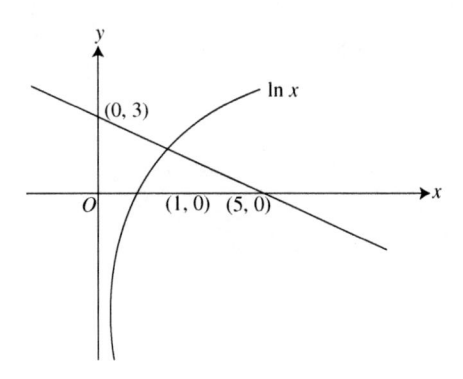

$$\frac{x}{5} + \frac{y}{3} = 1$$

$$y = 3\left(1 - \frac{x}{5}\right) = \ln x$$

$$1 - \frac{x}{5} = \frac{1}{3} \ln x$$

$$x = 5\left(1 - \frac{1}{3} \ln x\right)$$

$$x_{n+1} = 5\left(1 - \frac{1}{3} \ln x_n\right)$$

$$x_1 = 5\left(1 - \frac{1}{3} \ln 3\right) = 3.168979519$$

$$x_2 = 3.126389124$$

$$x_3 = 3.10020216$$

$$x_4 = 3.11422113.$$

---

5. $\sqrt{3} \cos x + \sqrt{2} \sin x \equiv R \cos(x - \alpha)$

$$\equiv R \cos x \cos \alpha + R \sin x \sin \alpha.$$

Equating the coefficients of $\cos x$ and $\sin x$, we have:

$\sqrt{3} = R \cos \alpha \quad \cdots (1)$

$\sqrt{2} = R \sin \alpha \quad \cdots (2)$

constructing the right angled triangle in Fig. 2

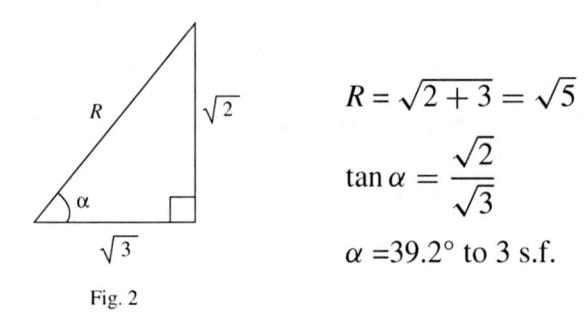

$R = \sqrt{2+3} = \sqrt{5}$

$\tan \alpha = \dfrac{\sqrt{2}}{\sqrt{3}}$

$\alpha = 39.2°$ to 3 s.f.

Fig. 2

$\therefore \sqrt{3}\cos x + \sqrt{2}\sin x \equiv \sqrt{5}\cos(x - 39.2°)$

$$\sqrt{5}\cos(x - 39.2°)$$

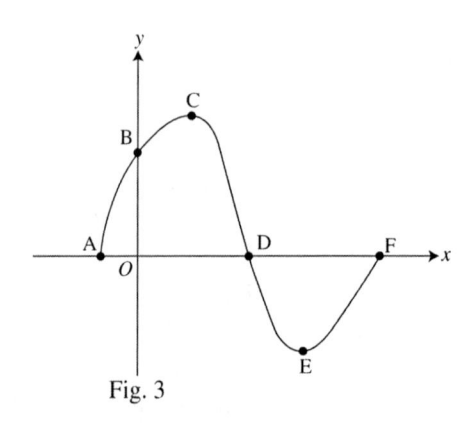

Fig. 3

$$x = 90° - 39.2° = 50.8°$$

A$(-50.8°, 0)$, B$(0, 0.775)$, C$(39.2°, \sqrt{5})$, D$(129.2°, 0)$
E$(219.2°, -\sqrt{5})$ and F$(309.2°, 0)$.

---

6. (a) $\sin(A - B) = \sin A \cos B - \sin B \cos A$

$\sin(60° - 45°) = \sin 15°$

$= \sin 60° \cos 45° - \sin 45° \cos 60°$

$= \dfrac{\sqrt{3}}{2} \dfrac{1}{\sqrt{2}} - \dfrac{1}{\sqrt{2}} \dfrac{1}{2} = \dfrac{\sqrt{3}}{2} \dfrac{\sqrt{2}}{2} - \dfrac{\sqrt{2}}{2} \dfrac{1}{2}$

$= \dfrac{\sqrt{2}}{4}(\sqrt{3} - 1).$

(b) $\cos 75° = \cos(45° + 30°)$

$= \cos 45° \cos 30° - \sin 45° \sin 30°$

$= \dfrac{1}{\sqrt{2}} \dfrac{\sqrt{3}}{2} - \dfrac{1}{\sqrt{2}} \dfrac{1}{2}$

$= \dfrac{1}{2\sqrt{2}} (\sqrt{3} - 1)$

$= \dfrac{\sqrt{2}}{4} (\sqrt{3} - 1)$

$\therefore \sin 15° = \cos 75°.$

(c) $\tan(45° - 30°) = \dfrac{\tan 45° - \tan 30°}{1 + \tan 45° \tan 30°}$

$= \dfrac{1 - \frac{\sqrt{3}}{3}}{1 + 1 \times \frac{\sqrt{3}}{3}} = \dfrac{1 - \frac{\sqrt{3}}{3}}{1 + \frac{\sqrt{3}}{3}}$

$= \dfrac{3 - \sqrt{3}}{3 + \sqrt{3}} \times \dfrac{3 - \sqrt{3}}{3 - \sqrt{3}} = \dfrac{(3 - \sqrt{3})^2}{9 - 3}$

$= \dfrac{9 + 3 - 6\sqrt{3}}{6} = \dfrac{12 - 6\sqrt{3}}{6}$

$= 2 - \sqrt{3}.$

(d) $\tan(A + B) = \dfrac{1}{\cot(A + B)}$

$\cot(A + B) = \dfrac{1}{\tan(A + B)} = \dfrac{1 - \tan A \tan B}{\tan A + \tan B}$

$= \dfrac{1 - \tan 30° \tan 45°}{\tan 30° + \tan 45°} = \left(1 - \dfrac{\sqrt{3}}{3} \times 1\right) \div \left(\dfrac{\sqrt{3}}{3} + 1\right).$

$= \dfrac{3 - \sqrt{3}}{3 + \sqrt{3}} \times \dfrac{(3 - \sqrt{3})}{(3 - \sqrt{3})} = \dfrac{(3 - \sqrt{3})^2}{9 - 3} = \dfrac{1}{6}\left(3 - \sqrt{3}\right)^2.$

$\cot 75° = \dfrac{9 + 3 - 6\sqrt{3}}{6} = 2 - \sqrt{3}$

$\therefore \tan 15° = \cot 75° = 2 - \sqrt{3}.$

7. (a) $f(x) = x^3\sqrt{2x^3 - 1}$

$$f'(x) = 3x^2\sqrt{2x^3 - 1} + x^3 \cdot \frac{1}{2}(2x^3 - 1)^{-\frac{1}{2}} \times 6x$$

$$= 3x^2\sqrt{2x^3 - 1} + 3x^4(2x^3 - 1)^{-\frac{1}{2}}$$

$$= 3x^2\sqrt{2x^3 - 1} + \frac{3x^4}{\sqrt{2x^3 - 1}}$$

$$= \frac{3x^2(2x^3 - 1) + 3x^4}{\sqrt{2x^3 - 1}} = \frac{6x^5 - 3x^2 + 3x^4}{\sqrt{2x^3 - 1}}$$

$$= \frac{3x^2(2x^3 - 1 + x^2)}{\sqrt{2x^3 - 1}}$$

$$= \frac{9x^4 - 3x}{\sqrt{2x^3 - 1}} = \frac{3x(3x^3 - x)}{\sqrt{2x^3 - 1}}.$$

(b) $g(x) = \dfrac{\cos 2x}{x^2 + 1}$

$$g'(x) = \frac{-2(x^2 + 1)\sin 2x - 2x \cos 2x}{(x^2 + 1)^2} = \frac{(-2x^2 - 2)\sin 2x - 2x \cos 2x}{(x^2 + 1)^2}$$

$$= \frac{-2\left[(x^2 + 1)\sin 2x + x \cos 2x\right]}{(x^2 + 1)^2}.$$

---

8. (i)

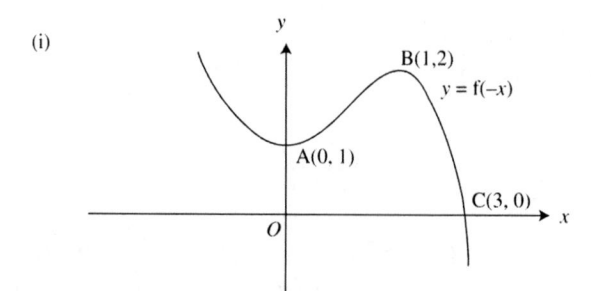

(ii)

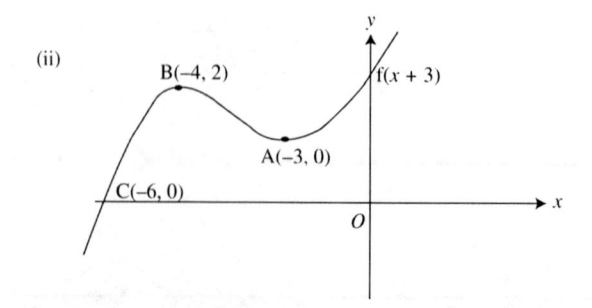

(iii)

Graph of $-f(x)$ showing points $C(-3, 0)$, $A(0, -1)$, and $B(-1, -2)$.

(iv)

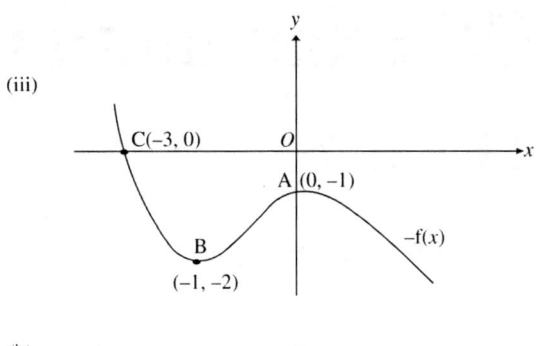

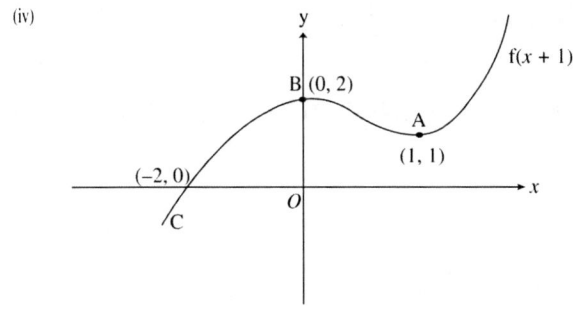

Graph of $f(x+1)$ showing points $B(0, 2)$, $A(1, 1)$, and $(-2, 0)$, with C below.

TOTAL FOR PAPER: 75 MARKS

GCE Examinations

Test Paper 7

Advanced Level

Core Mathematics C3

Time: 1 hour 30 minutes

Instructions and Information

Candidates may use any calculator allowed by the regulations of their Examination Board.

Full marks are awarded for correct answers to ALL questions.

This paper has eight questions.

You can start working with any question and you must label clearly all parts.

1. The function f is defined by $f: x \mapsto \frac{1}{x+1}$ where $x \in \mathbb{R}$ and $x \neq -1$.

   (a) Sketch this function, stating the equation of the asymptote. (3)

   (b) Determine $f^{-1}(x)$ and sketch it, stating the equation of the asymptote. (3)

   (c) State the domain and the range for $f(x)$ and $f^{-1}(x)$. (2)

   (d) Evaluate (i) $f^{-1}(1)$ (ii) fff (3). (3)

   (e) State whether $f(x)$ and $f^{-1}(x)$ are one-to-one function or many-to-one. (2)

---

2. The cubic function $f(x) = x^3 + 3x + 28$ is drawn on a graph paper and found that has only one solution approximately equal to $x_0 = -2.7$.

   If the iteration formula is used $x_{n+1} = -\frac{28}{x_n^2 + 3}$ we obtain

   $x_1 = -2.721$

   $x_2 = -2.691$

   $x_3 = -2.734$

   check these values. (3)

   If, however, we use the Newton-Raphson formula $x_{n+1} = x_n - \frac{f(x_n)}{f'(x_n)}$ we would obtain a better solution.

   Find $f(-2.7)$ and $f'(-2.7)$ and hence calculate x_{n+1}, if $x_n = -2.7$. (4)

---

3. A curve is given by the equation

   $$y = \frac{2\sin 3x + 1}{2\cos 3x - 1} = f(x).$$

   (a) Determine the first derivative. (4)

   (b) Determine the gradient at $x = \frac{\pi}{4}$ in the form $k\sqrt{2}$ where k is to be found. (6)

4. The decay graph relating n and t is shown in Fig. 1, where n is the number of nuclei at time t

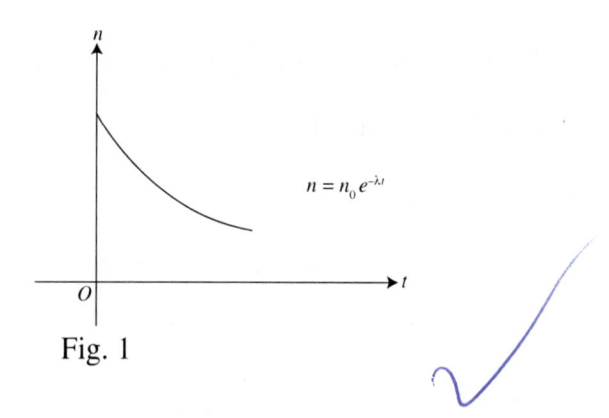

$n = n_0 e^{-\lambda t}$

Fig. 1

where $n = n_0$ at $t = 0$ and λ is a constant.

(a) Show that the rate of decay is $\dfrac{dn}{dt} = -\lambda n$. (2)

(b) Determine the time T required for n to fall from n_0 to $\tfrac{1}{2} n_0$. (4)

The half-life of certain substance is 5000 years find the decay constant λ per second. (2)

5. (a) Sketch the following trigonometric functions for $0 \le x \le 2\pi$.

(i) $f(x) = 2 \sin x$ (1)

(ii) $f(2x)$ (2)

(iii) $2f(2x)$ (2)

(iv) $f\left(\tfrac{1}{2}x\right) = 2 \sin \tfrac{x}{2}$. (2)

(b) Solve the simultaneous equation

(i) $2f(2x)$

(ii) $f\left(\dfrac{1}{2}x\right)$

by sketching these functions on the same graph and giving the approximate non-zero solutions to 2 significant figures. (3)

6. Solve the equation $\sin(3x - \frac{\pi}{4}) = -\frac{1}{2}$ for $0 \leq x \leq 2\pi$.

   The sketch of the graph is shown in Fig. 2 write down the coordinates of the point A, B, C, D, E, F, G, H, I, J and K.

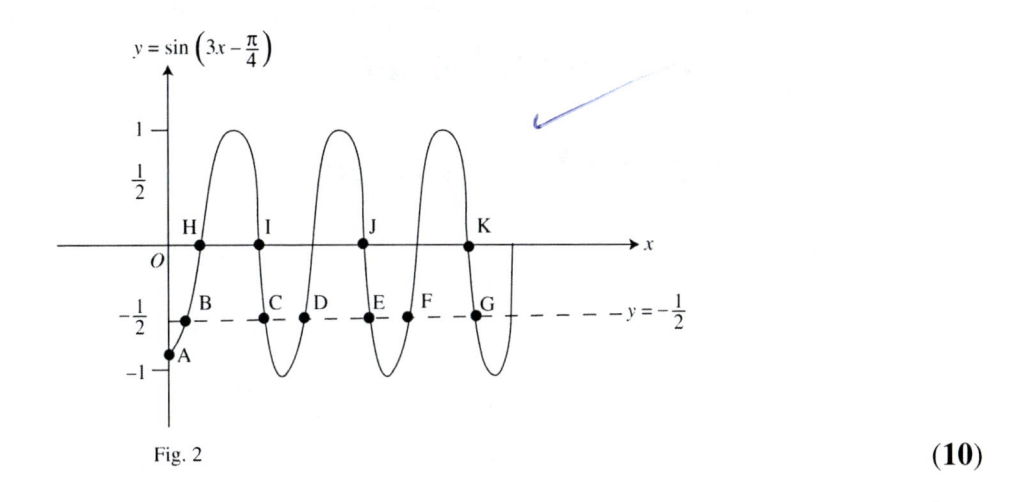

   Fig. 2 (10)

7. Find the derivatives for the following functions:

   (a) $f(x) = 5\cot^2 4x$ (3)

   (b) $g(x) = \ln(\sec x + \tan x)$ (3)

   (c) $h(x) = e^{6x} \cosec 4x$. (3)

8. Find the relationship between x and y if $\sin^{-1} x + \sin^{-1} y = \frac{\pi}{4}$ or $\theta + \phi = \frac{\pi}{4}$ where $\theta = \sin^{-1} x$ and $\phi = \sin^{-1} y$. (8)

TOTAL FOR PAPER: 75 MARKS

GCE Examinations

Test Paper 7 Solutions

Advanced Level

Core Mathematics C3

1. (a) $f: x \mapsto \dfrac{1}{x+1}$ has two asymptotes:

(i) when $\boxed{x = -1}$, $y \mapsto \infty$ and (ii) when $\boxed{y = 0}$, $x \mapsto \infty$.

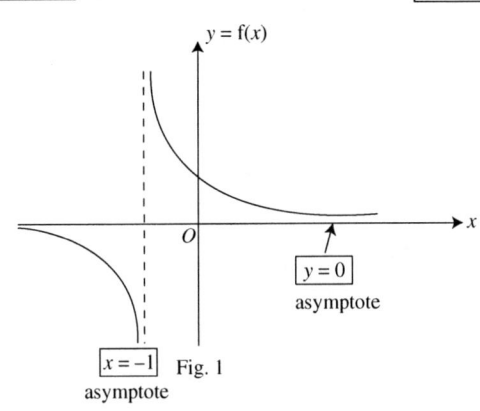

$y = f(x)$

$\boxed{y = 0}$ asymptote

$\boxed{x = -1}$ Fig. 1
asymptote

(b) $f(x) = y = \dfrac{1}{x+1}$, replace y for x and x for y and solve for y

$$x = \dfrac{1}{y+1} \Rightarrow y + 1 = \dfrac{1}{x} \Rightarrow y = \dfrac{1}{x} - 1$$

$\therefore f^{-1} = \dfrac{1}{x} - 1$.

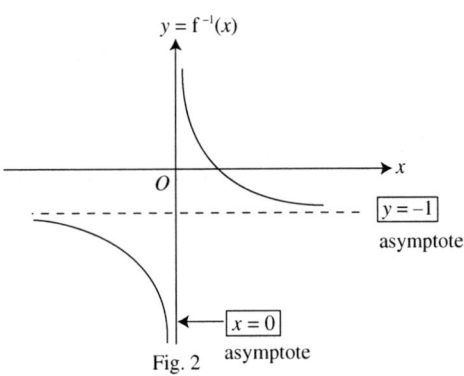

$y = f^{-1}(x)$

$\boxed{y = -1}$ asymptote

$\boxed{x = 0}$ asymptote
Fig. 2

(c) For Fig. 1 $x \in \mathbb{R}$ except for $\boxed{x = -1}$ and $y \in \mathbb{R}$ except for $\boxed{y = 0}$

For Fig. 2 $x \in \mathbb{R}$ except for $\boxed{x = 0}$ and $y \in \mathbb{R}$ except for $\boxed{y = -1}$

(d) (i) $f^{-1}(1) = \dfrac{1}{1} - 1 = 0$

(ii) $ff(x) = \dfrac{1}{\dfrac{1}{x+1} + 1} = \dfrac{x+1}{1+x+1} = \dfrac{x+1}{x+2}$

(iii) $fff(x) = \dfrac{\frac{1}{x+1}+1}{\frac{1}{x+1}+2} = \dfrac{1+x+1}{1+2x+2} = \dfrac{x+2}{2x+3}$

$\therefore fff(3) = \dfrac{5}{9}.$

(e) $f(x)$ and $f^{-1}(x)$ are one-to-one functions.

---

2. $x_{n+1} = -\dfrac{28}{x_n^2+3} \Rightarrow x_1 = -\dfrac{28}{(-2.7)^2+3} = -2.721088435$

$= -2.721$ to 3 d.p.

$x_2 = -\dfrac{28}{(-2.721088435)^2+3} = -2.691189226 = -2.691$ to 3 d.p.

$x_3 = -\dfrac{28}{(-2.691189226)^2+3} = -2.733707737 = -2.734$ to 3 d.p.

$f(x_n) = f(-2.7) = (-2.7)^3 + 3(-27) + 28 = 0.217$

$f'(x) = f'(-2.7) = 3(-2.7)^2 + 3 = 24.87$

$x_{n+1} = x_n - \dfrac{f(x_n)}{f'(x_n)} = -2.7 - \dfrac{0.217}{24.87} = -2.708125372$

$= -2.709.$ to 3 d.p.

---

3. $y = \dfrac{2\sin 3x + 1}{2\cos 3x - 1}$

$\dfrac{dy}{dx} = \dfrac{6\cos 3x(2\cos 3x - 1) - (2\sin 3x + 1)(-6\sin 3x)}{(2\cos 3x - 1)^2}$

$= \dfrac{12\cos^2 3x - 6\cos 3x + 12\sin^2 3x + 6\sin 3x}{(2\cos 3x - 1)^2}$

$= \dfrac{12(\cos^2 3x + \sin^2 3x) - 6(\cos 3x - \sin 3x)}{(2\cos 3x - 1)^2}$

$= \dfrac{12 - 6(\cos 3x - \sin 3x)}{(2\cos 3x - 1)^2}.$

Find the equation of the tangent at a point $x = \frac{\pi}{4}$.

$$y = mx + c$$

$$\frac{dy}{dx} = \frac{12 - 6\left(\cos\frac{3\pi}{4} - \sin\frac{3\pi}{4}\right)}{\left(2\cos\frac{3\pi}{4} - 1\right)^2}$$

$$= \frac{12 - 6\left(-\frac{1}{\sqrt{2}} - \frac{1}{\sqrt{2}}\right)}{\left[2\left(-\frac{1}{\sqrt{2}}\right) - 1\right]^2} = \frac{12 + \frac{12}{\sqrt{2}}}{\left(\frac{2}{\sqrt{2}} + 1\right)^2}$$

$$= \frac{12 + \frac{12}{\sqrt{2}} \cdot \frac{\sqrt{2}}{\sqrt{2}}}{\left(\frac{2\sqrt{2}}{2} + 1\right)^2} = \frac{12 + 6\sqrt{2}}{2 + 1 + 2\sqrt{2}} = \frac{12 + 6\sqrt{2}}{3 + 2\sqrt{2}} \times \frac{3 - 2\sqrt{2}}{3 - 2\sqrt{2}}$$

$$= \frac{36 + 18\sqrt{2} - 24\sqrt{2} - 24}{9 - 8} = 12 - 6\sqrt{2}.$$

$$y = mx + c = (12 - 6\sqrt{2})x + c$$

When $x = \frac{\pi}{4}$ $y = \dfrac{2\sin\frac{3\pi}{4} + 1}{2\cos\frac{3\pi}{4} - 1} = \dfrac{\frac{2}{\sqrt{2}} + 1}{-\frac{2}{\sqrt{2}} - 1} = -\dfrac{(2 + \sqrt{2})}{(2 + \sqrt{2})} = -1$

$$1 = (12 - 6\sqrt{2})\frac{\pi}{4} + c \Rightarrow c = 1 - (12 - 6\sqrt{2})\frac{\pi}{4}$$

$$y = (12 - 6\sqrt{2})x - 11 + 6\sqrt{2}\,\frac{\pi}{4}.$$

---

4. a) $n = n_0 e^{-\lambda t}$ $\dfrac{dn}{dt} = -\lambda n_0 e^{-\lambda t} = -\lambda n.$

b) If $n = n_0$ at $t = 0$ and let $t = T$ when $n = \dfrac{n_0}{2}$

$$\frac{n_0}{2} = n_0 e^{-\lambda t} \Rightarrow \frac{1}{2} = e^{-\lambda t} \Rightarrow e^{\lambda T} = 2$$

$$\lambda T = \ln 2 \Rightarrow T = \frac{\ln 2}{\lambda}.$$

$$\lambda = \frac{\ln 2}{5000} = 1.38629 \times 10^{-4} \text{ per year} = \frac{1.38929 \times 10^{-4}}{365 \times 24 \times 3600} \text{ per second}$$

$$= 4.40 \times 10^{-12} \text{ s}^{-1} \text{ to 3 s.f.}$$

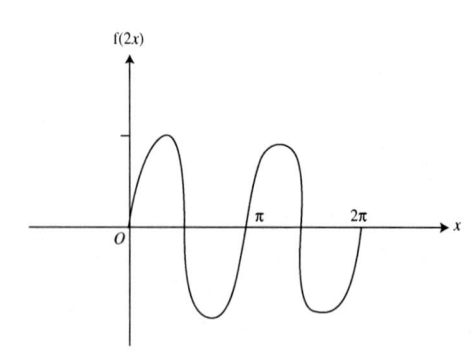

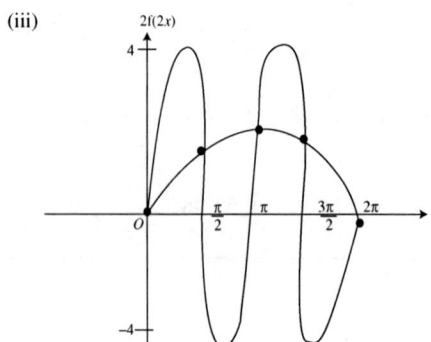

5. (a)

(i) *f(x)*

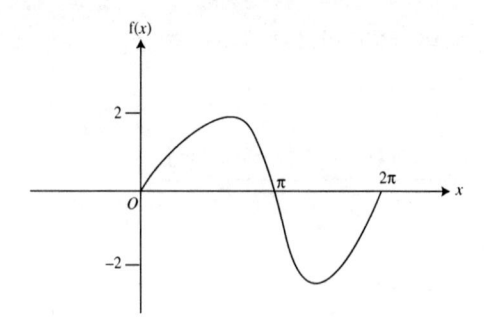

(ii) *f(2x)*

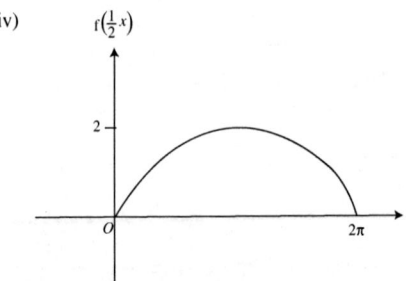

(iii) *2f(2x)*

(iv) $f(\tfrac{1}{2}x)$

(b) The three non-zero solutions are approximately $x < \dfrac{\pi}{2} \Rightarrow x = 1.5^c$

$$x > \pi \Rightarrow x = 3.2^c$$

$$x < \dfrac{3\pi}{2} \Rightarrow x = 4.5^c.$$

6. $\sin(3x - \dfrac{\pi}{4}) = -\dfrac{1}{2} = \sin\dfrac{7\pi}{6} = \sin\dfrac{11\pi}{6}$

$$= \sin\dfrac{19\pi}{6} = \sin\dfrac{23\pi}{6} = \sin\dfrac{31\pi}{6} = \sin\left(-\dfrac{\pi}{6}\right)$$

$3x - \dfrac{\pi}{4} = \sin\dfrac{7\pi}{6} \Rightarrow 3x = \dfrac{7\pi}{6} + \dfrac{\pi}{4} = \dfrac{14\pi}{12} + \dfrac{3\pi}{12} = \dfrac{17\pi}{12} \Rightarrow \boxed{x = \dfrac{17\pi}{36}}$

$3x - \dfrac{\pi}{4} = \dfrac{11\pi}{6} \Rightarrow 3x = \dfrac{\pi}{4} + \dfrac{11\pi}{6} = \dfrac{3\pi}{12} + \dfrac{22\pi}{12} = \dfrac{25\pi}{12}$ $\boxed{x = \dfrac{25\pi}{36}}$

$3x - \dfrac{\pi}{4} = \dfrac{19\pi}{6} \Rightarrow 3x = \dfrac{\pi}{4} + \dfrac{19\pi}{6} = \dfrac{3\pi}{12} + \dfrac{38\pi}{12}$ $\boxed{x = \dfrac{41\pi}{36}}$

$3x - \dfrac{\pi}{4} = \dfrac{23\pi}{6} \Rightarrow 3x = \dfrac{\pi}{4} + \dfrac{23\pi}{6} = \dfrac{3\pi}{12} + \dfrac{46\pi}{12}$ $\boxed{x = \dfrac{49\pi}{36}}$

$3x - \dfrac{\pi}{4} = \dfrac{31\pi}{6} \Rightarrow 3x = \dfrac{\pi}{4} + \dfrac{31\pi}{6} = \dfrac{3\pi}{12} + \dfrac{62\pi}{12}$ $\boxed{x = \dfrac{65\pi}{36}}$

$3x - \dfrac{\pi}{4} = -\dfrac{\pi}{6} \Rightarrow 3x = \dfrac{\pi}{4} - \dfrac{\pi}{6} = \dfrac{3\pi}{12} - \dfrac{2\pi}{12} = \dfrac{\pi}{12}$ $\boxed{x = \dfrac{\pi}{36}}$

$\therefore x = \dfrac{\pi}{36}, \dfrac{17\pi}{36}, \dfrac{25\pi}{36}, \dfrac{41\pi}{36}, \dfrac{49\pi}{36}, \dfrac{65\pi}{36}.$

A $\left(0, -\dfrac{1}{\sqrt{2}}\right)$ B $\left(\dfrac{\pi}{36}, -\dfrac{1}{2}\right)$ C $\left(\dfrac{17\pi}{36}, -\dfrac{1}{2}\right)$ D $\left(\dfrac{25\pi}{36}, -\dfrac{1}{2}\right)$ E $\left(\dfrac{41\pi}{36}, -\dfrac{1}{2}\right)$

F $\left(\dfrac{49\pi}{36}, -\dfrac{1}{2}\right)$ G $\left(\dfrac{65\pi}{36}, -\dfrac{1}{2}\right)$ H $\left(\dfrac{\pi}{12}, 0\right)$ I $\left(\dfrac{5\pi}{12}, 0\right)$ J $\left(\dfrac{13\pi}{12}, 0\right)$

K $\left(\dfrac{21\pi}{12}, 0\right)$

7. (a) $f(x) = 5\cot^2 4x$

$f'(x) = 5 \times 2\cot 4x \times 4\cot 4x \operatorname{cosec} 4x$

$\quad = 40\cot^2 4x \operatorname{cosec} 4x$

(b) $g(x) = \ln(\sec x + \tan x)$

$g'(x) = \dfrac{1}{\sec x + \tan x} \times \sec x \tan x + \sec^2 x$

$\quad = \dfrac{\sec x(\tan x + \sec x)}{\sec x + \tan x} = \sec x$

(c) $h(x) = e^{6x}\operatorname{cosec} 4x$

$h'(x) = 6e^{6x}\operatorname{cosec} 4x + e^{6x}4(-\operatorname{cosec} 4x \cot 4x)$

$\quad = 2e^{6x}\operatorname{cosec} 4x(3 - 2\cot 4x).$

---

8. $\sin^{-1} + \sin^{-1} = \dfrac{\pi}{4}$

taking sines on both sides of equation

$\sin(\sin^{-1} x + \sin^{-1} y) = \sin\dfrac{\pi}{4}$

$\sin(\sin^{-1} x)\cos(\sin^{-1} y) + \sin(\sin^{-1} y)\cos(\sin^{-1} x) = \dfrac{1}{\sqrt{2}}$

$x\cos(\sin^{-1} y) + y\sin^{-1} x = \dfrac{1}{\sqrt{2}} \quad \cdots (1)$

where $\sin\sin^{-1} x = x$ and $\sin\sin^{-1} y = y$ but $\sin^{-1} x = \theta \Rightarrow \sin\theta = x$
similarly $\sin(\sin^{-1} y) = y$.

From equation (1)

$$x\cos\phi + y\sin\theta = \dfrac{1}{\sqrt{2}}$$

$$x\sqrt{1 - \sin^2\phi} + y = \sqrt{1 - \sin^2\theta} = \dfrac{1}{\sqrt{2}}$$

$$\boxed{x\sqrt{1 - y^2} + y\sqrt{1 - x^2} = \dfrac{1}{\sqrt{2}}}$$

TOTAL FOR PAPER: 75 MARKS

GCE Examinations

Test Paper 8

Advanced Level

Core Mathematics C3

Time: 1 hour 30 minutes

Instructions and Information

Candidates may use any calculator allowed by the regulations of their Examination Board.

Full marks are awarded for correct answers to ALL questions.

This paper has eight questions.

You can start working with any question and you must label clearly all parts.

1. (a) Sketch the graph of $y = |x + 1|$, showing the coordinates of the points where the graph meets the coordinates axes. (4)

   (b) Sketch $y = -\dfrac{1}{x}$ and solve the simultaneous equations
   $$y = -\dfrac{1}{x} \qquad y = |x + 1|.$$ (4)

   Solve $x|x + 1| + 1 = 0$. (2)

2. Using $\sin(A + B) = \sin A \cos B + \sin B \cos A$

   show that $\sin 3x = 3 \sin x - 4 \sin^2 x$ and hence

   solve $\quad 6 \sin x - 8 \sin^3 x = 1 \quad$ for $0° \le x \le 360°$. (8)

3. (a) Show by calculation that the equation $x^3 + 3x + 3 = 0$ has only one real root and prove that this root lies between -0.7 and -0.9. (4)

   (b) Use the iteration formula $\quad x_{n+1} = -\dfrac{3}{x_n^2 + 3}$

   to find this root correct to 2 decimal places, take $x_0 = -0.8$ (6)

4. (a) Express $f(\theta) = 2 \sin \theta + 3 \cos \theta$ in the form of $R \sin(\theta + \alpha)$, where α is an acute angle. (4)

   (b) Hence solve the equation $f(\theta) = 1$. (4)

   (c) Sketch the graph $f(\theta)$, indicating on the graph the coordinate points of the maximum and minimum. (2)

5. Find the set of real values of x for which $|x + 4| > 2|x - 3|$. (6)

6.

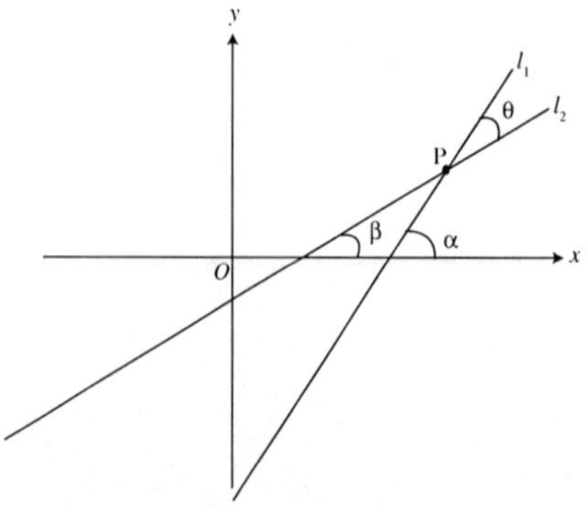

Fig. 1

(a) Fig. 1 shows two lines intersecting at P, the lines l_1 and l_2 make angles α and β respectively with the x-axis.

Show that $\tan \theta = \dfrac{m_1 - m_2}{1 + m_1 m_2}$

where $\tan \alpha = m_1$ and $\tan \beta = m_2$. (5)

Hence show that if $\theta = 0$, l_1 and l_2 are parallel and if $\theta = 90°$, l_1 and l_2 are perpendicular. (2)

(b) Find the equations of the tangent and normal at the point $(1, 9)$ of the curve $y = \dfrac{9}{x}$ giving your answer in the general form $ax + by + c = 0$. (5)

---

7. The functions f and g are defined for all the real values of x by $f(x) = 3x(x - 2)$ and $g(x) = mx + c$, where m and c are non-zero constants.

   (i) Find the range of f. (2)

   (ii) Given that $g(x) = g^{-1}(x)$ for all real values of x, show that $m = -1$. (4)

---

8. AC is a flexible ladder resting on a support BE and against a wall CD as shown in Fig. 2, the angle of AC with the horizontal floor AD is θ, the support is 2.7 high and is 6.4 m away from the wall.

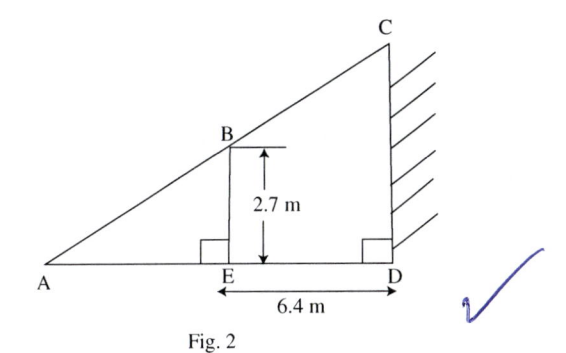

Fig. 2

Show that (i) $y = AC = \dfrac{2.7}{\sin \theta} + \dfrac{6.4}{\cos \theta}$ (4)

(ii) $\tan \theta = \dfrac{3}{4}$ assuming y is minimum (5)

Hence determine the minimum length of the ladder, y. (4)

TOTAL FOR PAPER: 75 MARKS

GCE Examinations

Test Paper 8 Solutions

Advanced Level

Core Mathematics C3

1. (a)

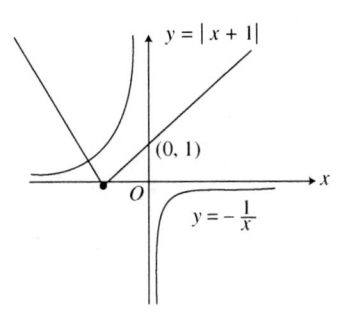

(b) $|x+1| = -\dfrac{1}{x} \Rightarrow x|x+1| + 1 = 0$

squaring up both sides

(c) $(x+1)^2 = \dfrac{1}{x^2}$

$\left(x+1-\dfrac{1}{x}\right)\left(x+1+\dfrac{1}{x}\right)$

$(x^2+x-1)(x^2+x+1) = 0$

$x^2+x+1 = 0$ has no real roots

$x^2+x-1 = 0$

$x = \dfrac{-1 \pm \sqrt{5}}{2}$ but the only solution

is $x = \dfrac{-1-\sqrt{5}}{2} = -1.618$.

---

2. $\sin(2x+x) = \sin 2x \cos x + \sin x \cos 2x$

$= 2\sin x \cos x \cos x + \sin x(1-2\sin^2 x)$

$= 2\sin x \cos^2 x + \sin x - 2\sin^3 x$

$= 2\sin x(1-\sin^2 x) + \sin x - 2\sin^3 x$

$= 2\sin x - 2\sin^3 x + \sin x - 2\sin^3 x$

$\sin 3x = 3\sin x - 4\sin^3 x$

$6\sin x - 8\sin^3 x = 2(3\sin x - 4\sin^3 x) = 2\sin 3x = 1$

$\sin 3x = \dfrac{1}{2} = \sin 30° = \sin 150° = \sin 390° = \sin 510°$

$= \sin 750° = \sin 870°$

$$3x = 30° \Rightarrow x = 10°$$
$$3x = 150° \Rightarrow x = 50°$$
$$3x = 390° \Rightarrow x = 130°$$
$$3x = 510° \Rightarrow x = 170°$$
$$3x = 750° \Rightarrow x = 250°$$
$$3x = 870° \Rightarrow x = 290°$$

$\therefore x = 10°, 50°, 130°, 170°, 250°, 290°$.

---

3. (i) $f(x) = x^3 + 3x + 3 = 0$

$f(-0.7) = (-0.7)^3 + 3(-0.7) + 3 = -0.343 - 2.1 + 3 = 0.557$

$f(-0.9) = (-0.9)^3 + 3(-0.9) + 3 = -0.429$,

change of sign and therefore $-0.9 \leq x \leq -0.7$.

(ii) $x_{n+1} = -\dfrac{3}{x_n^2 + 3}$

$x_0 = -0.8$

$x_1 = -\dfrac{3}{(-0.8)^2 + 3} = -0.824175824$

$x_2 = -\dfrac{3}{(-0.824175824)^2 + 3} = -0.81538007$

$x_3 = -\dfrac{3}{(x_2)^2 + 3} = -0.8185885583$

$x_4 = -\dfrac{3}{(x_3)^2 + 3} = -0.817419254$ $\boxed{\therefore x = -0.82}$

---

4. (a) $2\sin\theta + 3\cos\theta \equiv R\sin(\theta + \alpha) \equiv R\sin\theta\cos\alpha + R\cos\theta\sin\alpha$.

Equating the coefficients of $\sin\theta$ and $\cos\theta$ we have:

$2 = R\cos\alpha$ and $3 = R\sin\alpha$ and constructing the right angled triangle

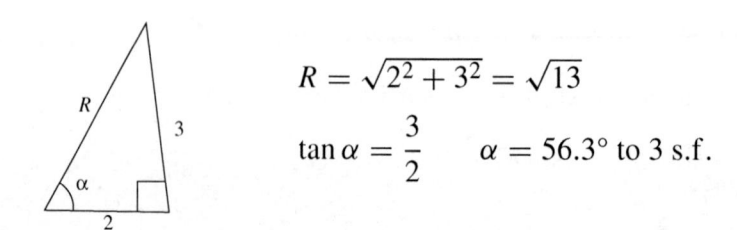

$R = \sqrt{2^2 + 3^2} = \sqrt{13}$

$\tan\alpha = \dfrac{3}{2}$ $\alpha = 56.3°$ to 3 s.f.

(b) $2\sin\theta + 3\cos\theta \equiv \sqrt{13}\sin(\theta + 56.3°)$

$\sqrt{13}\sin(\theta + 56.3°) = 1$

$\sin(\theta + 56.3°) = \frac{1}{\sqrt{13}} = \sin 16.1°$

$\theta + 56.3° = 16.1°$

$\theta = 16.1° - 56.3° = -40.2°.$

(c) $f(\theta) = \sqrt{13}\sin(\theta + 56.3°)$

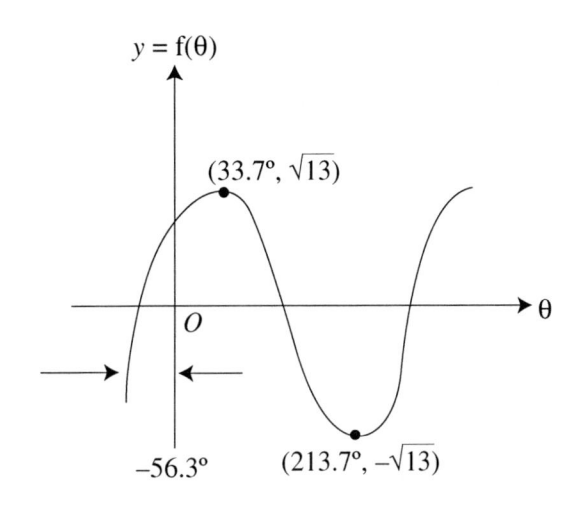

5. $$|x + 4| \geq 2|x - 3|$$

squaring up both sides

$$(x + 4)^2 \geq 2^2(x - 3)^2$$

$$[x + 4 - 2(x - 3)][x + 4 + 2(x - 3)] \geq 0$$

$$(-x + 10)(3x - 2) \geq 0$$

$$(x - 10)(3x - 2) \leq 0$$

$$\therefore \frac{2}{3} \leq x \leq 10.$$

6. (a) $\beta + \theta + 180° - \alpha = 180°$ $\boxed{\theta = \alpha - \beta}$

$$\tan\theta = \tan(\alpha - \beta)$$

$$= \frac{\tan\alpha - \tan\beta}{1 + \tan\alpha\tan\beta}$$

if $\theta = 0$, $\tan\theta = 0$

$\tan\alpha = \tan\beta$

$m_1 = m_2$

if $\theta = 90°$ $\tan 90° = \infty$

$1 + \tan\alpha\tan\beta = 0$ $\boxed{1 + m_1 m_2 = 0}$

$\therefore \boxed{m_1 m_2 = -1}$

(b) $y = \dfrac{9}{x} = 9x^{-1}$

$\dfrac{dy}{dx} = -9x^{-2} = -\dfrac{9}{x^2}$

$m_1 = -9$ when $x = 1$

$y = -9x + c_1$

$9 = -9x + c_1$

$c_1 = 18$

$y = -9x + 18 \Rightarrow \boxed{y + 9x - 18 = 0}$ the equation of the tangent

$m_2 = \dfrac{1}{9}$

$y = \dfrac{1}{9}x + c_2$

$9 = \dfrac{1}{9} + c_2 = 8\dfrac{8}{9} = \dfrac{80}{9}$

$y = \dfrac{1}{9}x + \dfrac{80}{9} \Rightarrow \boxed{9y - x - 80 = 0}$ the equation of the normal.

7. (i)

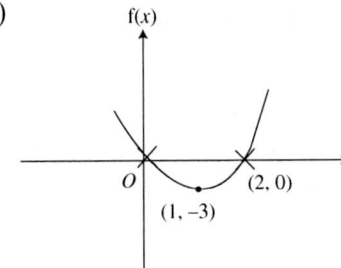

$f(x) = 3x^2 - 6x = 3x(x-2)$

$f'(x) = 6x - 6 = 0$

range ≥ -3

Graph of $f(x)$ parabola passing through O and $(2, 0)$ with minimum at $(1, -3)$.

(ii)
$$y = mx + c$$
$$x = my + c$$
$$y = \frac{x-c}{m}$$
$$g^{-1}(\alpha) = \frac{x-c}{m} = mx + c$$
$$x - c = m^2 x + mc \quad \text{equating the coefficients}$$
$$m^2 = 1 \text{ and } m_1 = -1 \quad \boxed{m = -1}$$

---

8. (i)

Fig. 2

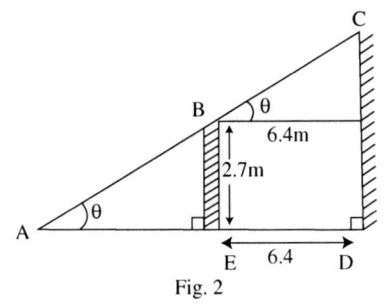

$0 < \theta < \frac{\pi}{2} \quad ABC = y = AC.$

To show $y = \frac{2.7}{\sin\theta} + \frac{6.4}{\cos\theta}$

$$\sin\theta = \frac{2.7}{AB} \Rightarrow AB = \frac{2.7}{\sin\theta}$$

$$\cos\theta = \frac{6.4}{BC} \Rightarrow BC = \frac{6.4}{\cos\theta}$$

$\therefore y = AB + BC = \frac{2.7}{\sin\theta} + \frac{6.4}{\cos\theta} = 2.7\,\text{cosec}\,\theta + 6.4\sec\theta$

(ii) $\dfrac{dy}{d\theta} = -2.7\operatorname{cosec}\theta \ \cot\theta + 6.4\sec\theta \tan\theta$ $\dfrac{dy}{d\theta} = 0$ for turning points

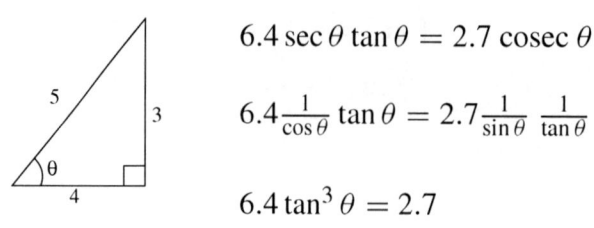

$6.4\sec\theta \tan\theta = 2.7\operatorname{cosec}\theta \ \cot\theta$

$6.4\dfrac{1}{\cos\theta}\tan\theta = 2.7\dfrac{1}{\sin\theta}\dfrac{1}{\tan\theta}$

$6.4\tan^3\theta = 2.7$

$\tan\theta = \sqrt[3]{\dfrac{27}{64}} = \dfrac{3}{4}$

$y_{min} = \dfrac{2.7}{\frac{3}{5}} + \dfrac{6.4}{\frac{4}{5}} = \dfrac{2.7 \times 5}{3} + \dfrac{6.4 \times 5}{4} = 0.9 \times 5 + 1.6 \times 5$

$= 4.5 + 8 = 12.5$ m

TOTAL FOR PAPER: 75 MARKS

GCE Examinations

Test Paper 9

Advanced Level

Core Mathematics C3

Time: 1 hour 30 minutes

Instructions and Information

Candidates may use any calculator allowed by the regulations of their Examination Board.

Full marks are awarded for correct answers to ALL questions.

This paper has eight questions.

You can start working with any question and you must label clearly all parts.

1. (a) Fig. 1 shows the function $f(x) = \sin x$

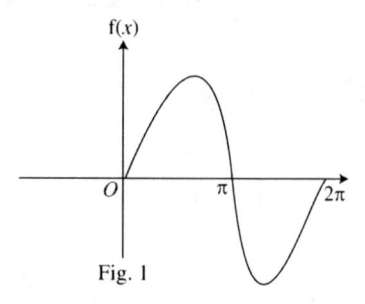

   Fig. 1

   Sketch $\frac{\pi}{2} - f(x)$. (2)

   (b) Fig. 2 shows the function $g(x) = x^3$

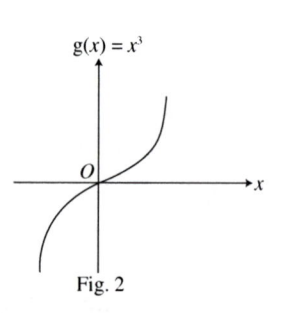

   Fig. 2

   Sketch $-g(x) + 2$. (2)

   (c) Fig. 3 shows the function $h(x) = \ln x$.

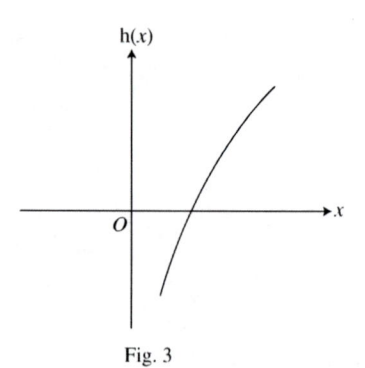

   Fig. 3

   Sketch $h^{-1}(x) + 2$. (2)

2. (a) Find the set of real values of x for which $|x - 1| \geq 3|2x - 3|$. (4)

   (b) Given that x satisfies the inequality in (a) find the greatest value of $|x + \frac{2}{5}|$. (2)

3. (a) If $\sin(A + B) = \sin A \cos B + \sin B \cos A$
   prove $\sin 4\theta = 2 \sin 2\theta \cos 2\theta$. (2)

   (b) If $\cos(A + B) = \cos A \cos B - \sin A \sin B$
   prove $\cos 4\theta = \cos^2 2\theta - \sin^2 2\theta = 2\cos^2 2\theta - 1 = 1 - 2\sin^2 2\theta$. (4)

   (c) Hence prove that $\tan 4\theta = \frac{2 \tan 2\theta}{1 - \tan^2 2\theta}$. (4)

---

4. (a) Show by calculation that the equation $e^{-x} - 2x = 0$ has a root between 0.34 and 0.36. (2)

   (b) Use the iteration formula $x_{n+1} = \frac{1}{2} e^{-x_n}$ with a suitable starting value, find this root correct to 4 decimal places. (4)

---

5. (a) C is the centre of the semicircle shown in Fig. 4. It is given that the area of the sector PCB is equal to the shaded segment. Show that $x = \frac{1}{2}(\pi - \sin x)$. (4)

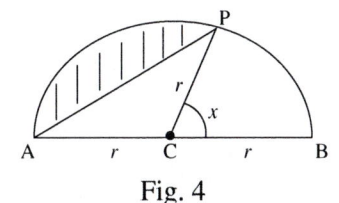

Fig. 4

   (b) $x_{n+1} = \frac{1}{2}(\pi - \sin x_n)$ is an iterative formula for evaluating x, starting with $x_0 = 1$, perform five iterations to find x_1, x_2, x_3, x_4, x_5 to 2 decimal places. (4)

   (c) Determine the area of the triangle ACP, if $r = 10$ cm and using the value for x_5, hence calculate the area of the segment to 2 significant figures. (4)

---

6. (a) If $f: x \mapsto 3x^{\frac{x}{2}} + 2$. Write the domain and range of the function and hence sketch the function indicating the coordinates of the point of intersection with the axis. (4)

   (b) Find the inverse function in a similar form. Write down the domain and range of this function and hence sketch the function indicating the coordinates of the point of intersection with the axis. (5)

---

7. (a) Find all the solutions $0 < x < 360°$ for $\tan^2 x - 3 \tan x + 2 = 0$ giving your answers to 1 decimal place. (4)

   (b) Prove that $\csc 2x + \cot 2x = \cot x$ and $\csc 2x - \cot 2x = \tan x$.
   Hence prove that $\csc^2 2x = 1 + \cot^2 2x$, and deduce the values of $\tan 22.5°$ and $\cot 22.5°$. (6)

8. Determine the derivative of the following logarithmic functions:

(a) $y = x \ln x^2$ (3)

(b) $y = x^2 \ln x$ (4)

(c) $y = x \ln x - x$ (4)

(d) $y = \sqrt{\dfrac{(x^2+1)}{(x^2-1)(x^4+1)}}$ show that

$\ln y = \frac{1}{2}\ln(x^2+1) - \frac{1}{2}\ln(x^2-1) - \frac{1}{2}\ln(x^4-1)$

hence show that $\dfrac{d}{dx}(\ln y) = \dfrac{x}{x^2+1} - \dfrac{x}{x^2-1} - \dfrac{2x^3}{x^4+1}$

and $\dfrac{dy}{dx} = y\left(\dfrac{x}{x^2+1} - \dfrac{x}{x^2-1} - \dfrac{2x^3}{x^4+1}\right).$ (5)

TOTAL FOR PAPER: 75 MARKS

GCE Examinations

Test Paper 9 Solutions

Advanced Level

Core Mathematics C3

1. (a)

$(0, \frac{\pi}{2})$

$\frac{\pi}{2} - f(x)$

(b)

$2 - g(x)$

$(0, 2)$

(c)

$h^{-1}(x) + 2$

e^x

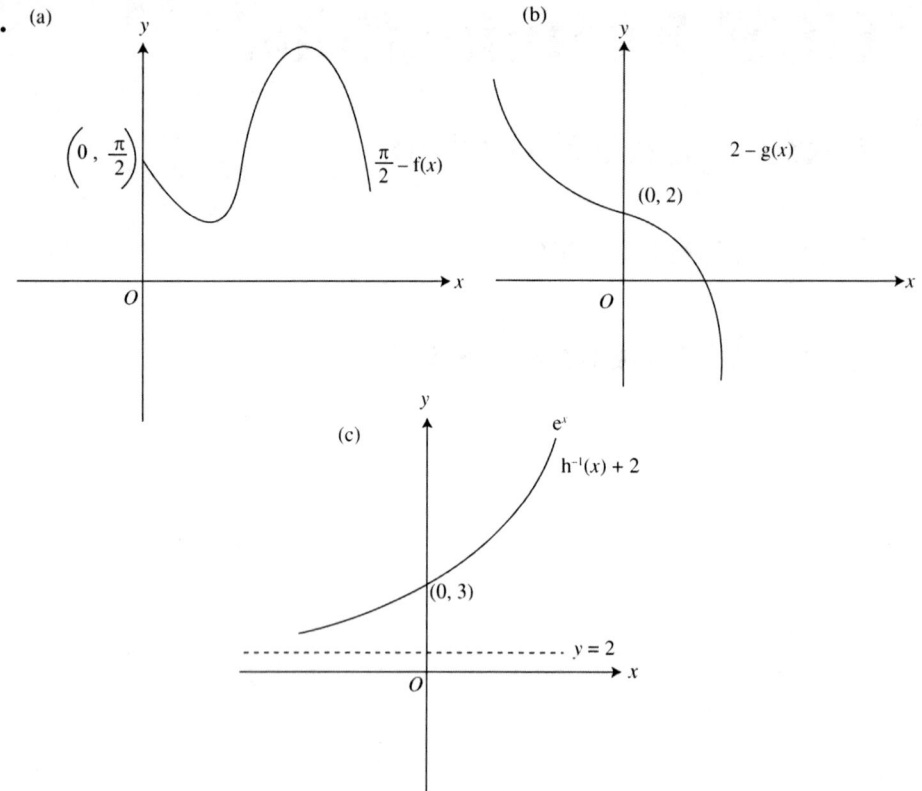

$(0, 3)$

$y = 2$

$$h(x) = \ln x$$
$$h^{-1}(x) = e^x.$$

---

2. (i) $|x - 1| \geq 3|2x - 3|$ squaring up both sides

$$(x - 1)^2 \geq 9(2x - 3)^2$$
$$[x - 1 - 3(2x - 3)][x - 1 + 3(2x - 3)] \geq 0$$
$$(x - 1 - 6x + 9)(x - 1 + 6x - 9) \geq 0$$
$$(-5x + 8)(7x - 10) \geq 0$$
$$(5x - 8)(7x - 10) \leq 0$$
$$\frac{10}{7} \leq x \leq \frac{8}{5}$$

or $\left\{x: \frac{10}{7} \leq x\right\} \cap \left\{x: \leq \frac{8}{5}\right\}$

(ii) The greatest possible value of $|x + \frac{2}{5}|$ is 2?

3. (a) $\sin(2\theta + 2\theta) = \sin 2\theta \cos 2\theta + \sin 2\theta \cos 2\theta$

$\therefore \sin 4\theta = 2\sin 2\theta \cos 2\theta$

(b) $\cos(2\theta + 2\theta) = \cos 2\theta \cos 2\theta - \sin\theta \sin\theta$

$\cos 4\theta = \cos^2 2\theta - \sin^2 2\theta$

$= \cos^2 2\theta - (1 - \cos^2 2\theta)$

$= 2\cos^2 2\theta - 1$

$= 1 - \sin^2 2\theta - \sin^2 2\theta = 1 - 2\sin^2 2\theta$

$\therefore \cos 4\theta = \cos^2 2\theta - \sin^2 2\theta = 2\cos^2 2\theta - 1 = 1 - 2\sin^2 2\theta.$

(c) $\tan 4\theta = \dfrac{\sin 4\theta}{\cos 4\theta} = \dfrac{2\sin 2\theta \cos 2\theta}{\cos^2 2\theta - \sin^2 2\theta}$

dividing numerator and denominator by $\cos^2 2\theta$ we have

$$\tan 4\theta = \dfrac{\frac{2\sin 2\theta \cos 2\theta}{\cos^2 2\theta}}{\frac{\cos^2 2\theta - \sin^2 2\theta}{\cos^2 2\theta}} = \dfrac{2\tan 2\theta}{1 - \tan^2 2\theta}.$$

---

4. (i) $f(x) = e^{-x} - 2x$

$f(0.34) = e^{-0.34} - 2(0.34)$

$= 0.711770322 - 0.68 = 0.031770322$

$f(0.36) = e^{-0.36} - 2 \times (0.36)$

$= 0.697676326 - 0.72$

$= -0.022323673$

$\therefore$ change of sign and $0.34 < x < 0.36$

(ii) Taking $x_0 = 0.35$

$x_1 = \frac{1}{2}e^{-0.35} = 0.352344044$

$x_2 = 0.351519101$

$x_3 = 0.351809204$

$x_4 = 0.351707158$

$x_5 = 0.35174305$

$\therefore x = 0.3517$

5. (a) The area of the sector PCB $= \pi r^2 \times \frac{x}{2\pi} = \frac{1}{2}r^2 x$.

The area of the $\triangle ACP = \frac{1}{2}r^2 \sin(\pi - x)$.

The shaded area $= \frac{\pi r^2}{2} - [\frac{1}{2}r^2 \sin(\pi - x) + \frac{1}{2}r^2 x] = \frac{1}{2}r^2 x$

$$\frac{\pi r^2}{2} - \frac{r^2}{2}\sin(\pi - x) - \frac{1}{2}r^2 x = \frac{1}{2}r^2 x$$

$$\frac{r^2}{2}[\pi - \sin(\pi - x)] = r^2 x$$

$$x = \frac{1}{2}(\pi - \sin(\pi - x))$$

$\sin(\pi - x) = \sin\pi \cos x - \sin x \cos\pi = \sin x$

$$\therefore x = \frac{1}{2}(\pi - \sin x^c).$$

(b) $x_2 = \frac{1}{2}(\pi - \sin x_1) = \frac{1}{2}(\pi - \sin 1^c)$

$\qquad = 1.150060834 = 1.15^c$

$x_3 = \frac{1}{2}(\pi - \sin 1.150060834^c)$

$\qquad = 1.114401933^c$

$x_3 = 1.11^c$ to 2 d.p.

$x_4 = \frac{1}{2}(\pi - \sin 1.114401933^c) = 1.121972641$

$x_4 = 1.12^c$

$x_5 = \frac{1}{2}(\pi - \sin 1.121972641^c)$

$\qquad = 1.120317255 \Rightarrow x_5 = 1.12 \qquad \therefore \boxed{x = 1.12^c}$

(c) Area $\triangle$ ACP $= \frac{1}{2}r^2 \sin(\pi - 1.12^c)$

$\qquad = \frac{1}{2}10^2 \sin(\pi - 1.12^c)$

$\qquad = 50 \times 0.9 = 45$ cm^2 to 2 s.f.

Area of $\triangle ACP = 45$ cm^2 $\quad$ 2 × Area segment + 45 = area of semi circle

$\qquad = \frac{\pi 10^2}{2} = 50\pi$.

Area segment $= \frac{50\pi - 45}{2} = 25\pi - 22.5 = 56.03981639 = 56$ cm^2.

6. (a) $f: x \mapsto 3e^{\frac{x}{2}} + 2$.

   $f:(0) = 3 + 2 = 5$

   $f(-\infty) \to 2$

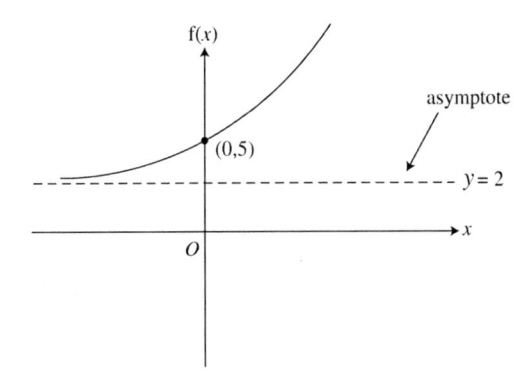

   The domain is for all real values of x, $x \in \mathbb{R}$. The range is $y > 2$.

   (b) $y = 3e^{\frac{x}{2}} + 2$

   replacing y and x respectively and solve for the new value of y

   $x = 3e^{\frac{y}{2}} + 2$

   $x - 2 = 3e^{\frac{y}{2}}$

   $3e^{\frac{y}{2}} = x - 2$

   $e^{\frac{y}{2}} = \frac{x-2}{3}$

   $\frac{y}{2} = \ln \frac{x-2}{3}$

   $y = 2 \ln \frac{x-2}{3}$

   $f^{-1}(x) = 2 \ln \frac{x-2}{3}$

   $f^{-1}: x \mapsto = 2 \ln \frac{x-2}{3}$.

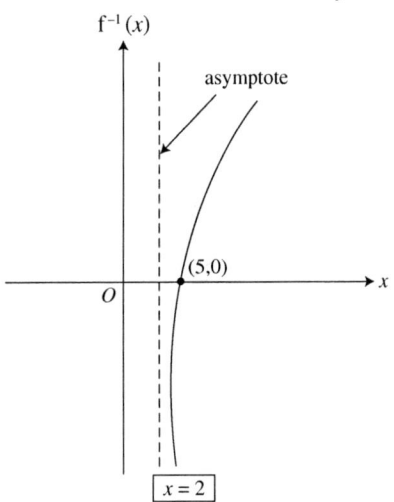

   The domain is when $x > 2$ and range is for all the real values, $y \in \mathbb{R}$.

   $f^{-1}(3) = 2 \ln \frac{3-2}{3} = 2 \ln \frac{1}{3} = -2.2$ to 2 s.f.

   $f^{-1}(5) = 2 \ln \frac{3}{3} = 2 \ln 1 = 0$.

7. (a) $\tan^2 x - 3 \tan x + 2 = 0$

   $(\tan x - 2)(\tan x - 1) = 0$

   $\tan x = 1 = \tan 45° = \tan 225°$

   $\tan x = 2 = \tan 63.4° = \tan 243.4°$

   $x = 45°, 63.4°, 225°, 243.4°$.

(b) $\csc 2x + \cot 2x = \frac{1+\tan^2 x}{2\tan x} + \frac{1-\tan^2 x}{2\tan x}$

$\qquad = \frac{2}{2\tan x} = \cot x.$

$\csc 2x - \cot 2x = \frac{1+\tan^2 x}{2\tan x} + \frac{1-\tan^2 x}{2\tan x}$

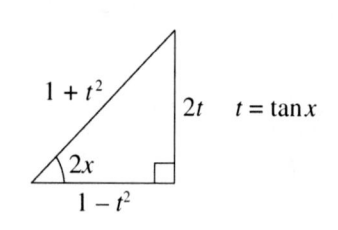

$\qquad = \frac{2\tan^2 x}{2\tan x} = \tan x$

(right triangle: hypotenuse $1+t^2$, opposite $2t$, adjacent $1-t^2$, angle $2x$, $t = \tan x$)

$(\csc 2x + \cot 2x)(\csc 2x - \cot 2x) = 1$

$\csc^2 2x - \cot^2 2x = 1.$

If $x = 22.5°$

$\tan x = \csc 2x - \cot 2x$

$\tan 22.5° = \csc 45° - \cot 45°$

$\qquad = \frac{\sqrt{2}}{1} - 1 = \sqrt{2} - 1$

$\cot 22.5° = \csc 45° + \cot 45°$

$\qquad = \frac{\sqrt{2}}{1} + 1 = \sqrt{2} + 1.$

---

8. (a) $y = x \ln x^2$

$\frac{dy}{dx} = 1 \cdot \ln x^2 + x \cdot \frac{1}{x^2} \cdot 2x = \ln x^2 + 2$

(b) $y = x^2 \ln x$

$\frac{dy}{dx} = 2x \ln x + x^2 \cdot \frac{1}{x} = 2x \ln x + x$

(c) $y = x \ln x - x$

$\frac{dy}{dx} = \ln x + x \cdot \frac{1}{x} - 1 = \ln x + 1 - 1 = \ln x$

(d) $y = \sqrt{\frac{x^2+1}{(x^2-1)(x^4+1)}} = \frac{(x^2+1)^{\frac{1}{2}}}{(x^2-1)^{\frac{1}{2}}(x^4+1)^{\frac{1}{2}}}$

$\ln y = \frac{1}{2}\ln(x^2+1) - \frac{1}{2}\ln(x^2-1) - \frac{1}{2}\ln(x^4+1)$

$\frac{1}{y}\frac{dy}{dx} = \frac{1}{2}\frac{2x}{x^2+1} - \frac{1}{2}\frac{2x}{x^2-1} - \frac{1}{2}\frac{4x^3}{x^4+1}$

$\frac{dy}{dx} = \sqrt{\frac{x^2+1}{(x^2-1)(x^4+1)}}\left(\frac{x}{x^2+1} - \frac{x}{x^2-1} - \frac{2x^3}{x^4+1}\right).$

TOTAL FOR PAPER: 75 MARKS

GCE Examinations

Test Paper 10

Advanced Level

Core Mathematics C3

Time: 1 hour 30 minutes

Instructions and Information

Candidates may use any calculator allowed by the regulations of their Examination Board.

Full marks are awarded for correct answers to ALL questions.

This paper has eight questions.

You can start working with any question and you must label clearly all parts.

1. (a) If $f: x \mapsto e^x + 2$ find $f^{-1}(x)$ and sketch both functions on the same graph. (3)
   (b) If $g: x \to \ln(x+3)$ find $g^{-1}(x)$ and sketch both functions on the same graph. (3)
   (c) $f(x) = e^{-x}$ and $g(x) = x + 2$. calculate $fg(1)$ and $gf(-1)$ to 3 decimal places. (3)

2. The function f is defined by

$$f: x \to \frac{1-x}{x+3}, \quad x \in \mathbb{R}, \quad x \neq -3.$$

   (a) Find $f^{-1}(x)$. (3)

   Fig. 1 shows the function $h(x)$ which has a domain $0 < x < 5$.

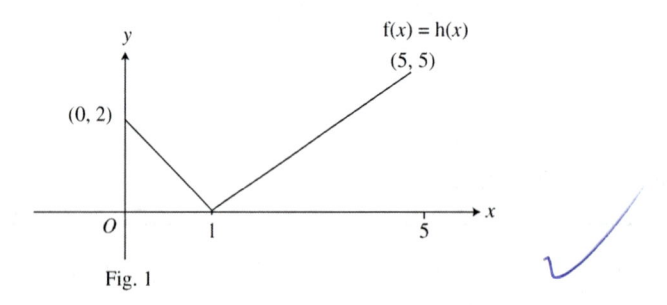

   Fig. 1

   (b) Write down the range of h. (1)
   (c) Find $hh(x)$, and hence work out $hh(4)$ state why $h(x)$ is <u>not</u> a one-to-one function. (3)
   (d) Sketch the inverse of $h(x)$ on separate diagram. Explain why the mapping is one-to-many. (3)
   (e) Write down the domain of $h^{-1}(x)$. (1)

3. (a) Express $f(x) = 3\sin x - 4\cos x$ in the form $R\sin(x - \alpha)$, where α is an acute angle. (3)
   (b) Hence solve $3\sin x - 4\cos x = 2$, giving your answer to 3 significant figures. (3)
   (c) Sketch the graph $f(x) = 3\sin x - 4\cos x$ for $0 \leq x \leq 360°$. (3)
   (d) Write down the coordinates of the maximum and minimum points of the graph (2)

4. (a) Show that $\sin x - \frac{1}{2}$ is a factor of $8\sin^3 x - 4\sin x + 1$ and show that this factorises to $(\sin x - \frac{1}{2})(8\sin^2 x + 4\sin x + 2)$. (4)
   (b) Solve $\cos x = \sin 4x$ for $0° \leq x \leq 360°$. (6)

5. (a) Find the set of real value of x for which $|-x+1| < |-x+2|$. (4)

   (b) Sketch the graphs $|-x+1|$ and $|-x+2|$ and verify the answer in (a). (4)

---

6. $f(x) = 2x^3 - 3x^2 - 11x + 8 = 0$.

   Show that one of the solutions of the cubic equation lies between -2.08 and -2.07. (4)

   Taking $x_0 = -2.075$ with an iterative formula of $f(x)$

   $$x_{n+1} = \frac{11x_n - 8}{2x_n^2 - 3x_n}$$

   find a solution to 5 decimal places. (4)

---

7. (a) Show that $\frac{d}{dx}(\ln x) = \frac{1}{x}$. (3)

   Hence find $\frac{d}{dx}(a^x)$ and write down $\frac{d}{dx}(2^x)$ and $\frac{d}{dx}(3^x)$. (3)

   (b) Find $\frac{d(\ln kx)}{dx}$ where k is a constant. (2)

---

8. A right circular cylinder is to be cut from a right circular cone, of height H and base of radius R. The axis of the cylinder lies along the axis of the cone. The circumference of one end of the cylinder is in contact with the curved surface of the cone and the other end of the cylinder lies on the base of the cone.

   Show that V, the volume of the cylinder is given by $V = \frac{\pi H x^2 (R-x)}{R}$, where x is the radius of the cylinder. (6)

   Show also that, as x varies, the maximum possible value of V is $\frac{4\pi R^2 H}{27}$. (4)

TOTAL FOR PAPER: 75 MARKS

GCE Examinations

Test Paper 10 Solutions

Advanced Level

Core Mathematics C3

1. (a) $y = e^x + 2$ replace x by y and y by x and solve for y
$x = e^y + 2 \Rightarrow e^y = x - 2 \Rightarrow y = \ln(x - 2)$
$f(x) = e^x + 2$ and $f^{-1}(x) = \ln(x - 2)$.

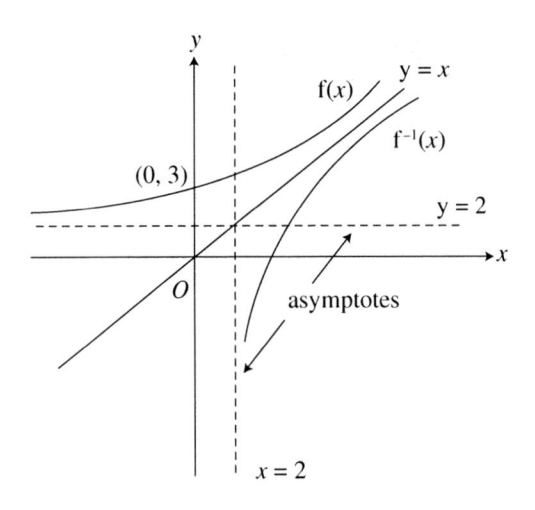

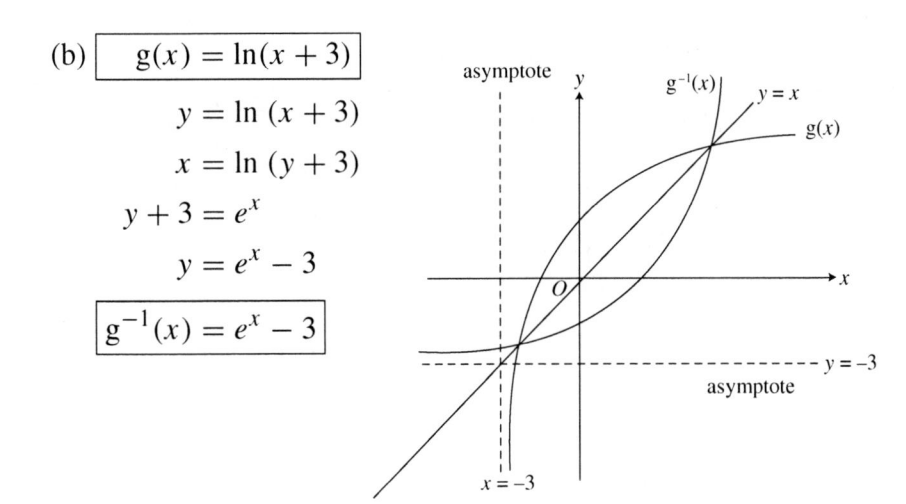

(b) $\boxed{g(x) = \ln(x + 3)}$

$y = \ln(x + 3)$
$x = \ln(y + 3)$
$y + 3 = e^x$
$y = e^x - 3$
$\boxed{g^{-1}(x) = e^x - 3}$

(c) $f(x) = e^{-x}$ and $g(x) = x + 2$

$fg(x) = e^{-(x+2)}$

$g(x) = x + 2$

$gf(x) = e^{-x} + 2$

$fg(1) = e^{-3} = 0.049787068 = 0.050$ to 3 d.p.

$gf(-1) = e + 2 = 4.718281828 = 4.718$ to 3 d.p.

2. (a) $f: x \mapsto \frac{1-x}{x+3}$ or $y = \frac{1-x}{x+3}$ replace x by y and y by x and solve for y

$$x = \frac{1-y}{y+3}$$

$$xy + 3x = 1 - y$$

$$xy + y = 1 - 3x$$

$$y = \frac{1 - 3x}{x + 1}$$

$$f^{-1}(x) = \frac{1 - 3x}{x + 1} \quad x \in \mathbb{R} \quad x \neq -1.$$

(b) The range $1 \leq y \leq 5$.

(c) $y = -2x + 2 \quad 0 \leq x \leq 1$

$y = \frac{5}{4}x - \frac{5}{4} \quad 1 \leq x \leq 5$

$\therefore h(x) = -2x + 2 \qquad hh(x) = -2(-2x + 2) + 2 = 4x - 2$

$h(x) = \frac{5}{4}x - \frac{5}{4} \qquad hh(x) = \frac{5}{4}\left(\frac{5}{4}x - \frac{5}{4}\right) - \frac{5}{4} = \frac{25}{16}x - \frac{25}{16} - \frac{20}{16}$

$$= \frac{25x}{16} - \frac{45}{16}$$

$hh(4) = 4(4) - 2 = 14$

$hh(4) = \frac{25}{16}(4) - \frac{45}{16} = \frac{55}{16}$

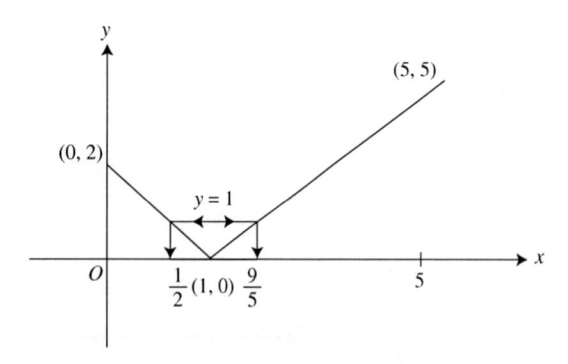

when $y = 1$ maps from two values of x

$1 = -2x + 2$

$-1 = -2x$

$x = \frac{1}{2}$ in $h(x) = -2x + 2$

$1 = \frac{5}{4}x - \frac{5}{4} \Rightarrow \frac{9}{4} = \frac{5}{4}x \Rightarrow x = \frac{9}{5}$

therefore $h(x)$ is a many-to-one function.

(d)

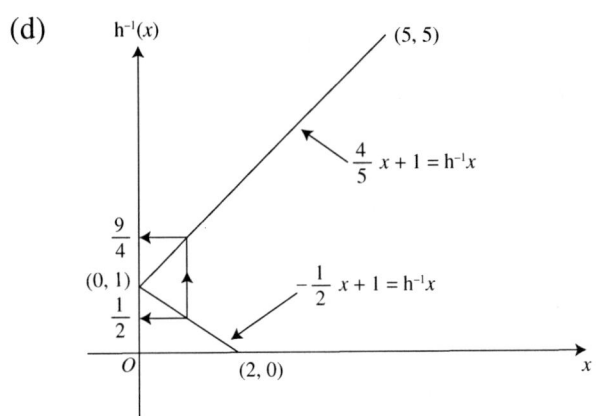

$y = -2x + 2$

$x = -2y + 2$

$2y = -x + 2$

$y = -\frac{1}{2}x + 1$

$h^{-1}(x) = -\frac{1}{2}x + 1$

$y = \frac{5}{4}x - \frac{5}{4}$

$x = \frac{5}{4}y - \frac{5}{4}$

$4x = 5y - 5$

$5y = 4x + 5$

$y = \frac{4}{5}x + 1$

$h^{-1}(x) = \frac{4}{5}x + 1.$

The mapping is one-to-many.

For the value of $x = 1$ the mapping has two values of y, $\frac{9}{4}$ and $\frac{1}{2}$.

(e) The domain for $h^{-1}(x)$ is $0 \leq x \leq 5$.

---

3. (a) $f(x) = 3\sin x - 4\cos x \equiv R\sin(x - \alpha)$

$\equiv R\sin x \cos\alpha - R\cos x \sin\alpha.$

Equating the coefficients of $\sin x$ and $\cos x$ we have:

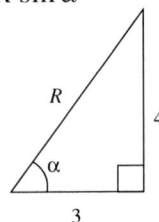

$3 = R\cos\alpha$ and $4 = R\sin\alpha$

$R = \sqrt{3^2 + 4^2} = 5$

$\sin\alpha = \frac{4}{5} = 0.8 \quad \cos\alpha = \frac{3}{5} = 0.6$ or $\tan\alpha = \frac{4}{3}$

$\alpha = 53.1°$ to 3 s.f.

$\therefore 3\sin x - 4\cos x \equiv 5\sin(x - 53.1°).$

(b) $5\sin(x - 53.1°) = 2$

$$\sin(x - 53.1°) = \frac{2}{5} = 0.4 = \sin 23.6°$$

$$x - 53.1° = 23.6$$

$$x = 76.7°.$$

(c)

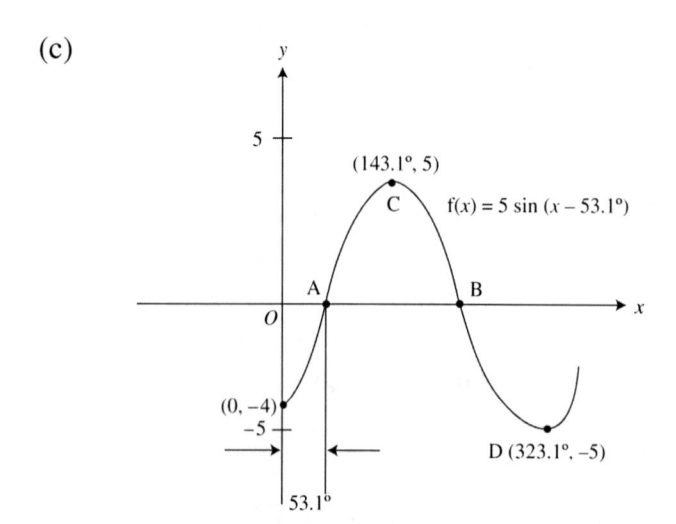

[Graph showing $f(x) = 5\sin(x - 53.1°)$ with points $(0, -4)$, A, B on x-axis, C$(143.1°, 5)$ maximum, and D$(323.1°, -5)$ minimum; phase shift of $53.1°$ indicated.]

(d) C$(143.1°, 5)$
D$(323.1°, -5)$.

---

4. (a) $f(\sin x) = 8\sin^3 x - 4\sin x + 1$

$$f\left(\frac{1}{2}\right) = 8\left(\frac{1}{2}\right)^3 - 4\left(\frac{1}{2}\right) + 1 = 1 - 4 \times \frac{1}{2} + 1 = 2 - 2 = 0$$

$$\left(\sin x - \frac{1}{2}\right)\left(8\sin^2 x + 4\sin x - 2\right) = 8\sin^3 x + 4\sin^2 x$$

$$- 2\sin x - 4\sin^2 x - 2\sin x + 1 = 8\sin^3 x - 4\sin x + 1$$

$$\therefore 8\sin^3 x - 4\sin x + 1 = \left(\sin x - \frac{1}{2}\right)\left(8\sin^2 + 4\sin x - 2\right).$$

(b) $\cos x = \sin 4x = 2\sin 2x \cos 2x = 2(2\sin x \cos x)(2\cos^2 x - 1)$

$$\therefore \cos x = 8\sin x \cos^3 x - 4\sin x \cos x$$

$$\cos x \left(1 - 8\cos^2 x \sin x + 4\sin x\right) = 0$$

$$\cos x = 0 \Rightarrow x = 0°, 360°$$

or $1 - 8\sin x(1 - \sin^2 x) + 4\sin x = 0 \Rightarrow 1 - 8\sin x + 8\sin^3 x + 4\sin x = 0$

$8\sin^3 x - 4\sin x + 1 = 0$

$\left(\sin x - \dfrac{1}{2}\right)\left(8\sin^2 x + 4\sin x - 2\right) = 0$

$\sin x = \dfrac{1}{2} \Rightarrow x = 30°, 150°$

$8\sin^2 x + 4\sin x - 2 = 0$

$\sin x = \dfrac{-4 \pm \sqrt{16 + 64}}{16} \Rightarrow \sin x = 0.3090169 \Rightarrow x = 18°$

or $\sin x = -0.90169 \Rightarrow x = 306°$.

---

5. (a)

$|-x + 1| < |-x + 2|$

squaring up both sides

$(-x + 1)^2 < (-x + 2)^2$

$\left[(-x + 1) - (-x + 2)\right]\left[(-x + 1) + (-x + 2)\right] < 0$

$(-x + 1 + x - 2)(-2x + 3) < 0$

$-1(-2x + 3) < 0$

$+2x - 3 < 0$

$2x < 3$

$x < \dfrac{3}{2}.$

(b)

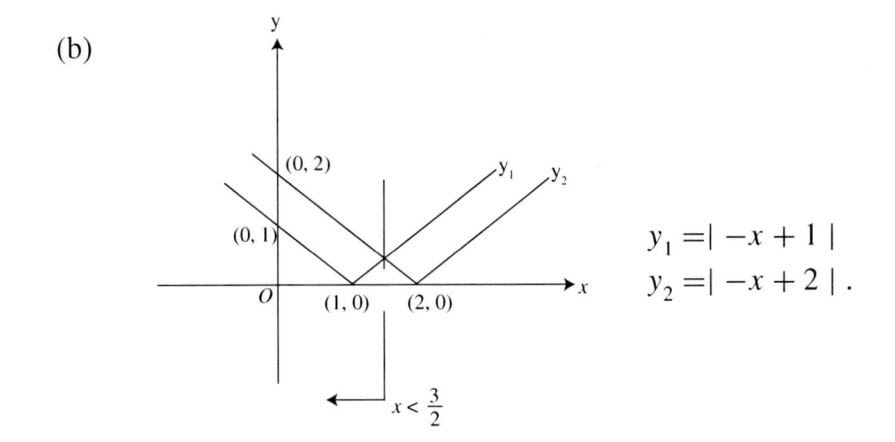

$y_1 = |-x + 1|$
$y_2 = |-x + 2|$.

6. $f(x) = 2x^3 - 3x^2 - 11x + 8 = 0$

$f(-2.08) = 2(-2.08)^3 - 3(-2.08)^2 - 11(-2.08) + 8 = -0.098$

$f(-2.07) = 2(-2.07)^3 - 3(-2.07)^2 - 11(-2.07) + 8 = 0.175814$

x lies between -2.08 and -2.07 because of change of sign

$2x^3 - 3x^2 - 11x + 8 = 0$

$x(2x^2 - 3x) = 11x - 8$

$$x_{n+1} = \frac{11x_n - 8}{2x_n^2 - 3x_n}$$

$x_0 = -2.075$

$$x_1 = \frac{11(-2.075) - 8}{2(-2.075)^2 - 3(-2.075)} = \frac{-30.825}{14.83625}$$

$x_1 = -2.077681355 = -2.07768.$

---

7. (a) if $y = \ln x$ then $e^y = x \quad \dfrac{dx}{dy} = e^y$

$$\frac{dy}{dx} = \frac{1}{e^y} = \frac{1}{x} \quad \therefore \frac{d}{dx}(\ln x) = \frac{1}{x}.$$

Let $y = a^x$

taking logs on both sides to the base e

$\log_e y = \log_e a^x = x \log_e a$

$\ln y = \ln a^x = x \ln a$

$\dfrac{d}{dx}(\ln y) = \dfrac{d}{dx}(x \ln a)$

$\dfrac{1}{y} \dfrac{dy}{dx} = \ln a \quad \dfrac{dy}{dx} = y \ln a = a^x \ln a$

$\dfrac{d}{dx}(a^x) = a^x \ln a.$

$\dfrac{d}{dx}(2^x) = 2^x \ln 2$

$\dfrac{d}{dx}(3^x) = 3^x \ln 3.$

(b) $y = \ln kx = \ln k + \ln x$

$$\frac{d}{dx}(\ln kx) = \frac{d}{dx}(\ln k + \ln x) = \frac{d}{dx}(\ln k) + \frac{d}{dx}(\ln x)$$

$$= 0 + \frac{1}{x} = \frac{1}{x}$$

Alternatively $\frac{d}{dx}(\ln kx) = \frac{1}{kx} \times k = \frac{1}{x}$ function of a function.

8.

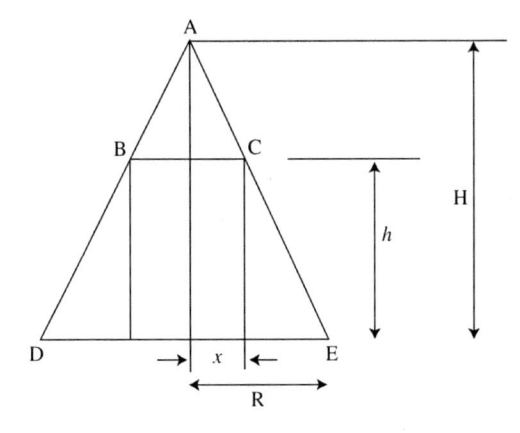

$V = \pi x^2 h$ $\triangle ABC$ and $\triangle ADE$ are similar $\frac{2x}{2R} = \frac{(H-h)}{H}$

$$\frac{x}{R} = 1 - \frac{h}{H} \Rightarrow \frac{h}{H} = 1 - \frac{x}{R} \Rightarrow h = \left(1 - \frac{x}{R}\right) H$$

$$V = \pi x^2 \left(1 - \frac{x}{R}\right) H = \frac{\pi H x^2 (R - x)}{R}$$

$$V = \pi H x^2 - \frac{\pi H x^3}{R}$$

$$\frac{dV}{dx} = 2\pi H x - \frac{3\pi H}{R} x^2.$$

For turning points $\frac{dV}{dx} = 0$

$$2\pi H = \frac{3\pi H}{R} x \Rightarrow x = \frac{2R}{3}$$

$$\frac{d^2V}{dx^2} = 2\pi H - \frac{6\pi H}{R}x$$

$$\frac{d^2V}{dx^2} = 2\pi H - \frac{6\pi H}{R}\left(\frac{2R}{3}\right) = -2\pi H < 0 \text{ maximum}$$

$$V_{\max} = \pi H \left(\frac{2R}{3}\right)^2 \left(R - \frac{2R}{3}\right)\frac{1}{R}.$$

$$V_{\max} = \pi H \frac{4R^2}{9}\left(1 - \frac{2}{3}\right)$$

$$= \frac{4\pi H R^2}{27}.$$

TOTAL FOR PAPER: 75 MARKS

GCE Examinations

Test Paper 1

Advanced Level

Core Mathematics C4

Time: 1 hour 30 minutes

Instructions and Information

Candidates may use any calculator allowed by the regulations of their Examination Board.

Full marks are awarded for correct answers to ALL questions.

This paper has seven questions.

You can start working with any question and you must label clearly all parts.

1. (a) Express $f(x) = \dfrac{16x^2+9}{16x^2-9}$ $x \neq \pm\dfrac{3}{4}$

   into partial fractions. (6)

   (b) Hence find the exact value of the definite integral $\displaystyle\int_1^2 \dfrac{16x^2+9}{16x^2-9}\,dx$. (6)

2. (a) If $y\tan x = y^2 x^2 + 3$ show that $\dfrac{dy}{dx} = \dfrac{y(2xy - \sec^2 x)}{\tan x - 2yx^2}$. (5)

   (b) If $x^2 y + y^2 x = 5$ find $\dfrac{dx}{dy}$. (3)

3. Sketch the line $y = 3x$ for $0 \leq x \leq h$. (2)

   A cone is now generated by the revolution of the line about the x-axis. Determine the volume of the cone so formed with radius r and height h. (6)

4. (a) Write down the expansion of $(1+x)^n$ where n is not a positive integer and state the constraint on x. (3)

   (b) Write down the first five terms in the expansion of (i) $\sqrt{1-2x}$ and (ii) $\sqrt{1+2x}$ and hence find $\sqrt{1-2x} + \sqrt{1+2x}$. (7)

5. A spherical balloon has a radius of 10 m. Air is pumped into the baloon at the rate of 0.005 m^3/s.

   Determine (i) the rate at which the radius increases (4)

   and (ii) the rate at which the surface area increases. (4)

6. The line l_1 has a vector equation $\mathbf{r} = \begin{pmatrix} 8 \\ -8 \\ 18 \end{pmatrix} + \lambda \begin{pmatrix} 2 \\ 2 \\ 6 \end{pmatrix}$,

   where λ is a parameter.

   The line l_2 has a vector equation

   $\mathbf{r} = \begin{pmatrix} 2 \\ 2 \\ 2 \end{pmatrix} + \mu \begin{pmatrix} 2 \\ -14 \\ 4 \end{pmatrix}$, where μ is a parameter.

   (a) Show that the lines intersect. (7)

   (b) Find the coordinates of their point of intersection. (3)

   (c) Given that θ is the angle between the lines, find $\cos\theta$. (7)

7. Evaluate approximately to three decimal places using the trapezoidal rule, the definite integral

$$\int_1^2 \log_{10} x \, dx$$

using 10 intervals or 11 ordinates.

x	1.0	1.1	1.2	1.3	1.4	1.5
$\log_{10} x$	0	.	.	0.0792	.	0.1761

x	1.6	1.7	1.8	1.9	2.0
$\log_{10} x$	.	0.2304	.	.	0.3010

(3)

use integration by parts to evaluate $\int_1^2 \log_e x \, dx$ and hence determine the integral $\int_1^2 \log_{10} x \, dx$.

(9)

TOTAL FOR PAPER: 75 MARKS

GCE Examinations

Test Paper 1 Solutions

Advanced Level

Core Mathematics C4

1. (a)
$$f(x) = \frac{16x^2+9}{16x^2-9} = \frac{16x^2-9+18}{16x^2-9} = 1 + \frac{18}{16x^2-9}$$

$$\frac{18}{16x^2-9} \equiv \frac{A}{4x-3} + \frac{B}{4x+3}$$

$$18 \equiv A(4x+3) + B(4x-3)$$

if $x = -\frac{3}{4}$, $18 = B(-6) \Rightarrow B = -3$,

if $x = \frac{3}{4}$, $18 = A(3+3) \Rightarrow A = 3$

$$\therefore \frac{18}{16x^2-9} \equiv \frac{3}{4x-3} - \frac{3}{4x+3}$$

$$f(x) = 1 + \frac{3}{4x-3} - \frac{3}{4x+3}.$$

(b)
$$\int_1^2 \frac{16x^2+9}{16x^2-9}\,dx = \int_1^2 \left(1 + \frac{3}{4x-3} - \frac{3}{4x+3}\right) dx$$

$$= \left[x + \frac{3}{4}\ln(4x-3) - \frac{3}{4}\ln(4x+3)\right]_1^2$$

$$= 2 + \frac{3}{4}\ln 5 - \frac{3}{4}\ln 11 - 1 - \frac{3}{4}\ln 1 + \frac{3}{4}\ln 7$$

$$= 1 + \frac{3}{4}\ln 5 + \frac{3}{4}\ln \frac{7}{11}$$

$$= 1 + \frac{3}{4}\ln \frac{35}{11}.$$

---

2. (a) $y\tan x = y^2 x^2 + 3$ differentiating with respect to x

$$\frac{dy}{dx}\tan x + y\sec^2 x \frac{dx}{dx} = 2y\frac{dy}{dx}x^2 + y^2 2x \frac{dx}{dx}$$

$$\frac{dy}{dx}(\tan x - 2yx^2) = 2xy^2 - y\sec^2 x$$

$$\frac{dy}{dx} = \frac{2xy^2 - y\sec^2 x}{\tan x - 2yx^2} = \frac{y(2x - \sec^2 x)}{\tan x - 2yx^2}$$

$$\therefore \frac{dy}{dx} = \frac{y(2x - \sec^2 x)}{\tan x - 2yx^2}.$$

(b)
$$x^2y + y^2x = 5 \text{ differentiating with respect to } y$$
$$2x\frac{dx}{dy}y + x^2 1 + 2yx + y^2\frac{dx}{dy} = 0$$
$$\frac{dx}{dy}(2xy + y^2) = -(x^2 + 2xy)$$
$$\frac{dx}{dy} = -\frac{x(x + 2y)}{y(2x + y)}.$$

3.

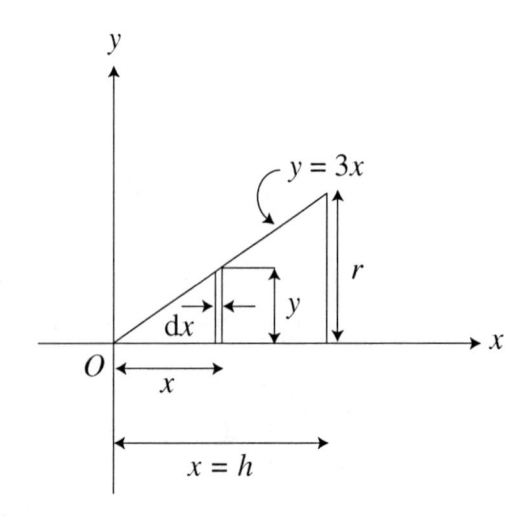

Determine the volume of the cone.

$$\pi y^2 dx = dV \text{ integrating both sides}$$
$$V = \int_0^h \pi y^2 dx$$
$$\text{but } \frac{y}{x} = \frac{r}{h} \Rightarrow y = \frac{rx}{h}$$
$$V = \int_0^h \pi \frac{r^2 x^2}{h^2} dx = \frac{\pi r^2}{h^2} \int_0^h x^2 dx$$
$$= \frac{\pi r^2}{h^2} \frac{1}{3} [x^3]_0^h$$
$$= \frac{1}{3}\pi r^2 h$$

the volume of the cone with radius r and height h.

4. (a) $(1+x)^n = 1 + \binom{n}{1}x + \binom{n}{2}x^2 + \binom{n}{3}x^3 + \ldots + \binom{n}{r}x^r + \ldots$

where $\binom{n}{r} = {}^nC_r = \dfrac{n!}{r!(n-r)!}$

$-1 < x < 1$ or $|x| < 1$.

(b) (i) $(1-2x)^{\frac{1}{2}} = 1 + \dfrac{1}{2}(-2x) + \dfrac{1}{2}\left(-\dfrac{1}{2}\right)(-2x)^2 \dfrac{1}{2!}$

$+ \left(\dfrac{1}{2}\right)\left(-\dfrac{1}{2}\right)\left(-\dfrac{3}{2}\right)(-2x)^3 \dfrac{1}{3!}$

$+ \left(\dfrac{1}{2}\right)\left(-\dfrac{1}{2}\right)\left(-\dfrac{3}{2}\right)\left(-\dfrac{5}{2}\right)(-2x)^4 \dfrac{1}{4!}$

$= 1 - x - \dfrac{1}{2}x^2 - \dfrac{1}{2}x^3 - \dfrac{5}{8}x^4.$

(ii) $(1+2x)^{\frac{1}{2}} = 1 + \dfrac{1}{2}(2x) + \dfrac{1}{2}\left(-\dfrac{1}{2}\right)(2x)^2\dfrac{1}{2!} + \left(\dfrac{1}{2}\right)\left(-\dfrac{1}{2}\right)\left(-\dfrac{3}{2}\right)(2x)^3\dfrac{1}{3!}$

$+ \left(\dfrac{1}{2}\right)\left(-\dfrac{1}{2}\right)\left(-\dfrac{3}{2}\right)\left(-\dfrac{5}{2}\right)(2x)^4\dfrac{1}{4!}$

$= 1 + x - \dfrac{1}{2}x^2 + \dfrac{1}{2}x^3 - \dfrac{5}{8}x^4$

$(1-2x)^{\frac{1}{2}} + (1+2x)^{\frac{1}{2}} = 2 - x^2 - \dfrac{5}{4}x^4.$

---

5. $V = \dfrac{4}{3}\pi\, 3r^2\, \dfrac{dr}{dt}$ ✓

(i) $\dfrac{dV}{dt} = \dfrac{4}{3}\pi\, 3r\, \dfrac{dr}{dt} \qquad 0.005 = 4\pi r^2 \dfrac{dr}{dt}$

$\therefore \dfrac{dr}{dt} = \dfrac{0.005}{4\pi\, 10^2} = \dfrac{0.00005}{4\pi} = 3.98 \times 10^{-6}\ \text{m/s}$

(ii) $S = 4\pi r^2$

$\dfrac{dS}{dt} = 4\pi\, 2r\, \dfrac{dr}{dt} = 8\pi r\, \dfrac{dr}{dt}$

$= 8\pi r \times \dfrac{0.00005}{4\pi} = 8\pi\, 10 \times \dfrac{0.00005}{4\pi} = 1 \times 10^{-3}\ \text{m}^2/\text{s}.$

6. (a) The direction ratios are 2:2:6 and 2:−14:4 or 1:1:3 and 1:−7:2 which are not equal therefore the lines are not parallel.

$$l_1: \quad \mathbf{r}_1 = \begin{pmatrix} 8 \\ -8 \\ 18 \end{pmatrix} + \lambda \begin{pmatrix} 2 \\ 2 \\ 6 \end{pmatrix}$$

$$l_2: \quad \mathbf{r}_1 = \begin{pmatrix} 2 \\ 2 \\ 2 \end{pmatrix} + \mu \begin{pmatrix} 2 \\ -14 \\ 4 \end{pmatrix}$$

$$\mathbf{r}_1 = \begin{pmatrix} x_1 \\ y_1 \\ z_1 \end{pmatrix} \quad \mathbf{r}_2 = \begin{pmatrix} x_2 \\ y_2 \\ z_2 \end{pmatrix}$$

Equating the coefficients of **i, j, k** for the two lines we have

$$8 + \lambda 2 = 2 + \mu 2 \Rightarrow \lambda = -3 + \mu \quad \ldots (1)$$
$$-8 + \lambda 2 = 2 + \mu(-14) \Rightarrow \lambda = 5 - 7\mu \quad \ldots (2)$$
$$8 + \lambda 6 = 2 + \mu 4 \Rightarrow 3\lambda = -8 + 2\mu \quad \ldots (3)$$

Solving (1) and (2)
$$-3 + \mu = 5 - 7\mu \Rightarrow \boxed{\mu = 1}$$

$$\lambda = -3 + 1 = -2 \Rightarrow \boxed{\lambda = -2}$$

Substituting $\mu = 1$ and $\lambda = -2$ in (3) $3(-2) = -8 + 2 = -6$, verifies these equations and the values of $\mu = 1$ and $\lambda = -2$ make the two vector equations equal, therefore the lines intersect.

(b) $\mathbf{r}_1 = \mathbf{r}_2$ and the position vector of the point of intersection is $(4\mathbf{i} - 12\mathbf{j} + 6\mathbf{k})$

$$\therefore (4, -12, 6).$$

(c) The angle between a pair of lines depends only on their directions and <u>not</u> on their positions, The direction vectors are

$$V_1 = 2\mathbf{i} + 2\mathbf{j} + 6\mathbf{k} \text{ and } V_2 = 2\mathbf{i} - 14\mathbf{j} + 4\mathbf{k}$$
$$V_1 \cdot V_2 = |V_1||V_2|\cos\theta$$
$$(2\mathbf{i} + 2\mathbf{j} + 6\mathbf{k}) \cdot (2\mathbf{i} - 14\mathbf{j} + 4\mathbf{k}) = 2 \times 2 - 28 + 24 = 0$$
$$\cos\theta = 0$$
$$\theta = \frac{\pi}{2}.$$

7. $\int_1^2 \log_{10} x \, dx = \dfrac{0.1}{2}[0 + 0.3010 + 2(0.0414 + 0.0792 + 0.1139 + 0.1461$
$\qquad\qquad\qquad\qquad + 0.1761 + 0.2041 + 0.2304 + 0.2553 + 0.2788)]$
$\qquad\qquad\qquad = 0.168$

where $h = \dfrac{2-1}{10} = 0.1$

x	1	1.1	1.2	1.3	1.4	1.5
$\log_{10} x$	0	0.0414	0.0792	0.11390	0.1461	0.1761

x	1.6	1.7	1.8	1.9	2.0
$\log_{10} x$	0.2041	0.2304	0.2553	0.2788	0.3010

$$\log_e x = \dfrac{\log_{10} x}{\log_{10} e}$$

$$\int_1^2 \log_{10} x \, dx = \log_{10} e \int_1^2 \log_e x \, dx$$

$$\int_1^2 \log_e x \, dx = \left[x \log_e x\right]_1^2 - \int_1^2 x \cdot \dfrac{1}{x} dx$$
$\qquad\qquad\qquad\quad ② \qquad\quad ①$
$$\qquad\qquad\quad = 2\ln 2 - [x]_1^2 \, dx$$
$$\qquad\qquad\quad = 2\ln 2 - 1$$
$$\qquad\qquad\quad = 0.386294361$$

$\therefore \int_1^2 \log_{10} x \, dx = \log_{10} e \times 0.386294361$
$\qquad\qquad\qquad\quad = 0.43429448 \times 0.386294361$
$\qquad\qquad\qquad\quad = 0.167765509$
$\qquad\qquad\qquad\quad = 0.168$ to 3 d.p.

TOTAL FOR PAPER: 75 MARKS

GCE Examinations

Test Paper 2

Advanced Level

Core Mathematics C4

Time: 1 hour 30 minutes

Instructions and Information

Candidates may use any calculator allowed by the regulations of their Examination Board.

Full marks are awarded for correct answers to ALL questions.

This paper has seven questions.

You can start working with any question and you must label clearly all parts.

1. (a) Express $\dfrac{6x^2 + 13x + 3}{x(x+1)(x+3)}$

   into partial fractions. (8)

   (b) Check your answers. (4)

---

2. Prove that the volume of a sphere of radius r is $V = \dfrac{4}{3}\pi r^3$. (6)

---

3. If $(1 - 2x)^{\frac{1}{2}} = 1 - x - \frac{1}{2}x^2$ and $(1 + 2x)^{\frac{1}{2}} = 1 + x - \dfrac{1}{2}x^2$

   by substituting $x = \dfrac{1}{20}$ and $x = \dfrac{1}{100}$ estimate $\sqrt{102}$ and $\sqrt{10}$
   respectively to three significant figures. (6)

---

4. A curve is given by the parametric equations $x = \theta - \sin\theta$ and $y = 1 - \cos\theta$.

   Fig. 1 shows the sketch of this curve for $0 \leq \theta \leq 2\pi$.

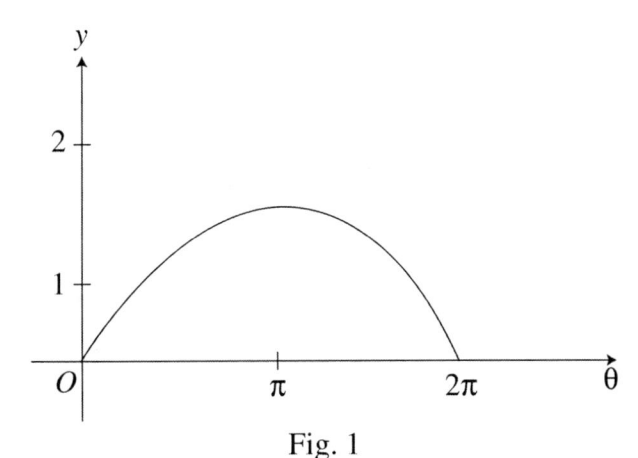

   Fig. 1

   (a) Show that $\dfrac{dy}{dx} = \cot\dfrac{\theta}{2}$. (6)

   (b) Determine the equations of the tangent and normal at $\theta = \dfrac{\pi}{2}$. (8)

---

5. An equilateral triangle has a side of 10 cm and its perimeter increases at the rate of 1 cm /s, find the rate of increase of the area. if each side increase to 10.1 cm, find the corresponding increase in the area. $A = \sqrt{s(s-a)(s-b)(s-c)}$ where $s = \dfrac{a+b+c}{2}$ is the Heron's formula for the area. (10)

6. With respect to an origin 0, the position vectors of the point P, Q an R are given as follow:
$$p = \overrightarrow{OP} = i + j + k$$
$$q = \overrightarrow{OQ} = 2i + 3j + 4k$$
$$r = \overrightarrow{OR} = -5i + 4j - 3k.$$ (4)

(a) Find the vectors $\overrightarrow{PQ}$ and $\overrightarrow{PR}$.
  Calculate the acute angle $\angle RPQ$
  (i) using the cosine rule (5)
  (ii) using the scalar product. (5)

7. Integrate the following integrals:

(a) $\displaystyle\int \frac{\sin x}{\sqrt{\cos x}}\, dx$ (4)

(b) $\displaystyle\int \frac{\ln x}{x}\, dx$ (4)

(c) $\displaystyle\int \frac{x}{\sqrt{7x^2 + 1}}\, dx.$ (5)

TOTAL FOR PAPER: 75 MARKS

GCE Examinations

Test Paper 2 Solutions

Advanced Level

Core Mathematics C4

1. (a) $\dfrac{6x^2 + 13x + 3}{x(x+1)(x+3)} \equiv \dfrac{A}{x} + \dfrac{B}{x+1} + \dfrac{C}{x+3}$

$6x^2 + 13x + 3 \equiv A(x+1)(x+3) + Bx(x+3) + Cx(x+1)$

if $x = -1$

$6(-1)^2 + 13(-1) + 3 = B(-1)(-1+3)$

$-4 = -2B$

$\boxed{B = 2}$

if $x = 0$

$6(0) + 13(0) + 3 = A(1)(3)$

$\boxed{A = 1}$

if $x = -3$

$6(-3)^2 + 13(-3) + 3 = C(-3)(-2)$

$54 - 39 + 3 = 6C$

$18 = 6C$

$\boxed{C = 3}$

$\dfrac{6x^2 + 13x + 3}{x(x+1)(x+3)} \equiv \dfrac{1}{x} + \dfrac{2}{x+1} + \dfrac{3}{x+3} \quad \cdots (1)$

(b) $\dfrac{1}{x} + \dfrac{2}{x+1} + \dfrac{3}{x+3} = \dfrac{x+1+2x}{x(x+1)} + \dfrac{3}{x+3} = \dfrac{3x+1}{x(x+1)} + \dfrac{3}{x+3}$

$= \dfrac{(1+3x)(x+3) + 3x(x+1)}{x(x+1)(x+3)}$

$= \dfrac{x + 3 + 3x^2 + 9x + 3x^2 + 3x}{x(x+1)(x+3)}$

$= \dfrac{6x^2 + 13x + 3}{x(x+1)(x+3)}.$

Therefore RHS = LHS in (1).

2.

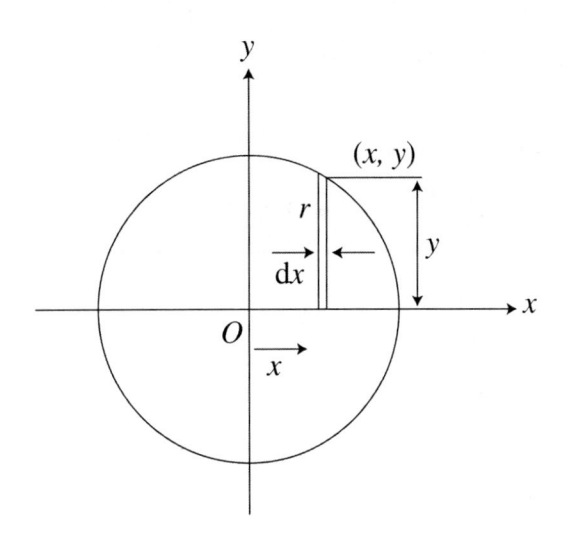

$dV = \pi y^2 \, dx \qquad x^2 + y^2 = r^2$

$V = \displaystyle\int_{-r}^{r} \pi y^2 \, dx \qquad y^2 = r^2 - x^2$

$= \pi \displaystyle\int_{-r}^{r} (r^2 - x^2) \, dx = \pi \left[r^2 x - \dfrac{x^3}{3} \right]_{-r}^{r}$

$= \pi \left[r^3 - \dfrac{r^3}{3} \right] - \left[-r^3 + \dfrac{r^3}{3} \right]$

$= \pi \left(\dfrac{2r^3}{3} + \dfrac{2r^3}{3} \right)$

$= \dfrac{4}{3}\pi r^3.$

---

3. $(1 - 2x)^{\frac{1}{2}} = 1 - x - \frac{1}{2}x^2$

$\left(1 - 2 \times \frac{1}{20}\right)^{\frac{1}{2}} = \left(1 - \frac{1}{10}\right)^{\frac{1}{2}} = \left(\frac{10-1}{10}\right)^{\frac{1}{2}} = \dfrac{9^{\frac{1}{2}}}{10^{\frac{1}{2}}}$

$= \dfrac{3}{\sqrt{10}} = 1 - \frac{1}{20} - \frac{1}{2} \times \frac{1}{20^2}$

$= 1 - 0.05 - 0.00125 = 0.94875$

$= \dfrac{3}{\sqrt{10}} \times \dfrac{\sqrt{10}}{\sqrt{10}} = \dfrac{3}{10}\sqrt{10} = 0.94875$

$\sqrt{10} = \dfrac{9.4875}{3} = 3.1625 = 3.16$ to 3 s.f.

$$(1+2x)^{\frac{1}{2}} = \left(1+2\left(\tfrac{1}{100}\right)\right)^{\frac{1}{2}} = \sqrt{\tfrac{102}{100}} = \tfrac{\sqrt{102}}{10}$$

$$= 1 + \tfrac{1}{100} - \tfrac{1}{2}\left(\tfrac{1}{100}\right)^2 = 1 + 0.01 - 0.00005 = 1.00995$$

$$\sqrt{102} = 10 \times 1.00995 = 10.1 \text{ to 3 s.f.}$$

---

4. (a) $x = \theta - \sin\theta \quad \dfrac{dx}{d\theta} = 1 - \cos\theta$

$y = 1 - \cos\theta \quad \dfrac{dy}{d\theta} = \sin\theta$

$$\frac{dy}{dx} = \frac{\frac{dy}{d\theta}}{\frac{dx}{d\theta}} = \frac{\sin\theta}{1-\cos\theta} = \frac{2\sin\frac{\theta}{2}\cos\frac{\theta}{2}}{1-(2\cos^2\frac{\theta}{2}-1)}$$

$$= \frac{2\sin\frac{\theta}{2}\cos\frac{\theta}{2}}{2-2\cos^2\frac{\theta}{2}} = \frac{2\sin\frac{\theta}{2}\cos\frac{\theta}{2}}{2(1-\cos^2\frac{\theta}{2})}$$

$$= \frac{\sin\frac{\theta}{2}\cos\frac{\theta}{2}}{\sin^2\frac{\theta}{2}} = \cot\frac{\theta}{2} \quad \therefore \frac{dy}{dx} = \cot\frac{\theta}{2}.$$

(b) Equation of the tangent $y = mx + c$

$m_1 = \dfrac{dy}{dx} = \cot\dfrac{\theta}{2}$

at $\theta = \dfrac{\pi}{2} \quad m = \cot\dfrac{\pi}{4} = 1$

$x = \dfrac{\pi}{2} - \sin\dfrac{\pi}{2} = \dfrac{\pi}{2} - 1$

$y = 1 - \cos\dfrac{\pi}{2} = 1$

$y = x + c \quad$ but $x = \dfrac{\pi}{2} - 1 \quad$ when $y = 1, \quad 1 = \dfrac{\pi}{2} - 1 + c \Rightarrow c = 2 - \dfrac{\pi}{2}$

$$\therefore \boxed{y = x + 2 - \dfrac{\pi}{2}}$$

Equation of the normal is $y = m_2 x + c$ but $m_1 m_2 = -1 \Rightarrow m_2 = -\dfrac{1}{m_1} = -\dfrac{1}{1} = -1$

$$y = -x + c$$

$$1 = -\left(\tfrac{\pi}{2} - 1\right) + c$$

$$c = 1 + \left(\tfrac{\pi}{2} - 1\right) = \tfrac{\pi}{2}$$

$$\boxed{y = -x + \dfrac{\pi}{2}}$$

the equation of the normal.

5. $A = \sqrt{s(s-a)(s-b)(s-c)} = \sqrt{15 \times 5 \times 5 \times 5} = 25\sqrt{3}$ cm^2

where $s = \frac{10+10+10}{2}$ 15 cm = the semi-perimeter.

$s = \frac{3}{2}a, \quad a = \frac{2}{3}s$

$A = \sqrt{s\left(s - \frac{2}{3}s\right)^3} = \frac{s^2}{3\sqrt{3}}$

$\frac{dA}{ds} = \frac{2s}{3\sqrt{3}} \Rightarrow \frac{dA}{dt} = \frac{2s}{3\sqrt{3}} \frac{ds}{dt} = \frac{2\times 15}{3\sqrt{3}} \times \frac{1}{2} = \frac{5}{3}\sqrt{3}$ cm^2/s

$A = \frac{1}{2}a^2 \sin 60° \quad \frac{dA}{da} = a \sin 60°$

$\delta A = a \sin 60° \, \delta a$

$= 10 \times 0.866 \times 0.1 = 0.866$ cm^2.

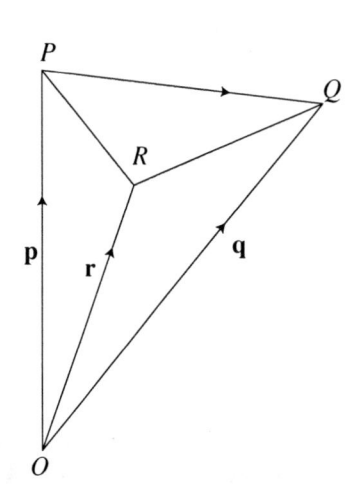

6.

$\vec{PQ} = \mathbf{q} - \mathbf{p} = 2\mathbf{i} + 3\mathbf{j} + 4\mathbf{k} - (\mathbf{i} + \mathbf{j} + \mathbf{k})$

$= \mathbf{i} + 2\mathbf{j} + 3\mathbf{k}$

$|\vec{PQ}| = \sqrt{1^2 + 2^2 + 3^2} = \sqrt{14}$

$\vec{PR} = \mathbf{r} - \mathbf{p} = -6\mathbf{i} + 3\mathbf{i} - 4\mathbf{k}$

$|\vec{PR}| = \sqrt{36 + 9 + 16} = \sqrt{61}$

$\vec{RQ} = \mathbf{q} - \mathbf{r} = 7\mathbf{i} - \mathbf{j} + 7\mathbf{k}$

$|\vec{RQ}| = \sqrt{49 + 1 + 49} = \sqrt{99}$

(b) $\angle RPQ$

(i) $RQ^2 = PR^2 + PQ^2 - 2(PR)(PQ)\cos \angle RPQ$

$\cos \angle RPQ = \frac{61 + 14 - 99}{2\sqrt{61}\sqrt{14}} = \frac{-24}{2\sqrt{61}\sqrt{14}} = -0.4106315$

$\angle RPQ = 114.24° \quad \therefore \angle RPQ = 65.76°$ the acute angle

(ii) $\vec{PR}.\vec{PQ} = (-6\mathbf{i} + 3\mathbf{j} - 4\mathbf{k}).(\mathbf{i} + 2\mathbf{j} + 3\mathbf{k}) = -6 + 6 - 12 = -12$

$= |\vec{PR}||\vec{PQ}| \cos \angle RPQ = \sqrt{61}\sqrt{14} \cos \angle RPQ$

$\therefore \angle RPQ = 65.76°$ the acute angle.

7. (a) $\displaystyle\int \frac{\sin x}{\sqrt{\cos x}}\,dx = \int \frac{d(-\cos x)}{(\cos x)^{\frac{1}{2}}}$

$\displaystyle\phantom{\int \frac{\sin x}{\sqrt{\cos x}}\,dx} = \int (\cos x)^{-\frac{1}{2}}\,d(-\cos x)$

$\displaystyle\phantom{\int \frac{\sin x}{\sqrt{\cos x}}\,dx} = -\frac{(\cos x)^{\frac{1}{2}}}{\frac{1}{2}} + c$

$\displaystyle\phantom{\int \frac{\sin x}{\sqrt{\cos x}}\,dx} = -2\sqrt{\cos x} + c$

(b) $\displaystyle\int \frac{\ln x}{x}\,dx = \int \frac{1}{x}\ln x\,dx = \int \ln x\,d(\ln x)$

$\displaystyle\phantom{\int \frac{\ln x}{x}\,dx} = \frac{(\ln x)^2}{2} + c$

(c) $\displaystyle\int \frac{x}{\sqrt{7x^2+1}}\,dx = \frac{1}{14}\int \frac{d(7x^2+1)}{(7x^2+1)^{\frac{1}{2}}}$

$\displaystyle\phantom{\int \frac{x}{\sqrt{7x^2+1}}\,dx} = \frac{1}{14}\int (7x^2+1)^{-\frac{1}{2}}\,d(7x^2+1)$

$\displaystyle\phantom{\int \frac{x}{\sqrt{7x^2+1}}\,dx} = \frac{1}{14}\cdot\frac{(7x^2+1)^{\frac{1}{2}}}{\frac{1}{2}} + c$

$\displaystyle\phantom{\int \frac{x}{\sqrt{7x^2+1}}\,dx} = \frac{1}{7}\sqrt{7x^2+1} + c.$

TOTAL FOR PAPER: 75 MARKS

GCE Examinations

Test Paper 3

Advanced Level

Core Mathematics C4

Time: 1 hour 30 minutes

Instructions and Information

Candidates may use any calculator allowed by the regulations of their Examination Board.

Full marks are awarded for correct answers to ALL questions.

This paper has seven questions.

You can start working with any question and you must label clearly all parts.

1. Express $f(x) = \frac{x^3 - 4x^2 - 18x + 23}{(x+2)(x-3)(x-5)}$ into partial fractions. (12)

2. (a) Sketch the graph $y = \frac{1}{x}$ in the first quadrant when $1 \leq x \leq 5$. (2)
   The curve is rotated about the x-axis through $360°$. (6)
   Determine the volume so generated.

   (b) $\int_0^3 \frac{x}{\sqrt{1+x^2}}\, dx$

   determine the exact value of this integral. (4)

3. The cycloid is given by the parametric equations
$$x = t - \sin t$$
$$y = 1 - \cos t.$$

   (a) Determine the value of $\frac{dy}{dx}$ in terms of t. (2)
   (b) Show that the cartesian equation of the curve is given by the equation (8)

4. (a) If y is small compared with x find the expansion of $(x+y)^{\frac{1}{2}}$ and
   hence show that $(x+y)^{\frac{1}{2}} \approx \sqrt{x} + \frac{1}{2}x^{-\frac{1}{2}}y$. (5)
   (b) Find correct to 4 decimal places the value of $\sqrt[3]{27.003}$. (5)

5. (a) Determine the define integral
$$\int_{-2}^{2} \frac{e^x}{1+e^x}\, dx$$

   using the trapezoidal rule with five ordinates. (5)

   (b) Find the exact definite integral
$$\int_{-2}^{2} \frac{e^x}{1+e^x}\, dx.$$
   (5)

6. The vector equations of two lines l_1 and l_2 are given:

   $l_1: \mathbf{r} = 2\mathbf{i} + 3\mathbf{j} + 5\mathbf{k} + \lambda(\mathbf{i} + \mathbf{j} + 2\mathbf{k})$

   $l_2: \mathbf{r} = 4\mathbf{j} + 6\mathbf{k} + \mu(-\mathbf{i} + 2\mathbf{j} + 3\mathbf{k})$

   where λ and μ are two parameters.

(a) Show that the lines intersect. (6)

(b) Find the coordinates of their point of intersection. (4)

(c) Given that θ is the acute angle between the lines, show that the exact value of $\cos\theta$ is $\sqrt{\frac{7}{12}}$. (6)

7. The area of a circle is given by $A = \pi r^2$.

   Find the relationship between the rate of change A and that of the radius r. (5)

TOTAL FOR PAPER: 75 MARKS

GCE Examinations

Test Paper 3 Solutions

Advanced Level

Core Mathematics C4

1. $\dfrac{x^3 - 4x^2 - 18x + 23}{(x+2)(x-3)(x-5)} = \dfrac{x^3 - 4x^2 - 18x + 23}{x^3 - 6x^2 - x + 30} = f(x)$

$(x+2)(x-3)(x-5) = (x^2 - x - 6)(x-5) = x^3 - x^2 - 6x - 5x^2 + 5x + 30$

$\qquad\qquad\qquad\qquad = x^3 - 6x^2 - x + 30$

$f(x) = \dfrac{x^3 - 6x^2 + 2x^2 - x - 17x + 30 - 7}{x^3 - 6x^2 - x + 30}$

$ = \dfrac{(x^3 - 6x^2 - x + 30) + (2x^2 - 17x - 7)}{x^3 - 6x^2 - x + 30}$

$ = 1 + \dfrac{2x^2 - 17x - 7}{(x+2)(x-3)(x-5)}$

$\dfrac{2x^2 - 17x - 7}{(x+2)(x-3)(x-5)} \equiv \dfrac{A}{x+2} + \dfrac{B}{x-3} + \dfrac{C}{x-5}$

$2x^2 - 17x - 7 \equiv A(x-3)(x-5) + B(x+2)(x-5) + C(x+2)(x-3)$

by the cover up rule if $x = -2$

$2(-2)^2 - 17(-2) - 7 = A(-5)(-7)$

$\qquad\qquad 35 = 35A \quad \boxed{A = 1}$

if $x = 3$

$2(3)^2 - 17(3) - 7 = B(5)(-2)$

$\qquad\qquad -40 = -10B \quad \boxed{B = 4}$

if $x = 5$

$2(5)^2 - 17(5) - 7 = C(7)(2)$

$\qquad\qquad 14C = -42 \quad \boxed{C = -3}$

$f(x) = \dfrac{1}{x+2} + \dfrac{4}{x-3} - \dfrac{3}{x-5} + 1$

$ = \dfrac{1}{x+2} + \dfrac{4}{x-3} + \dfrac{x-8}{x-5}.$

2. (a)

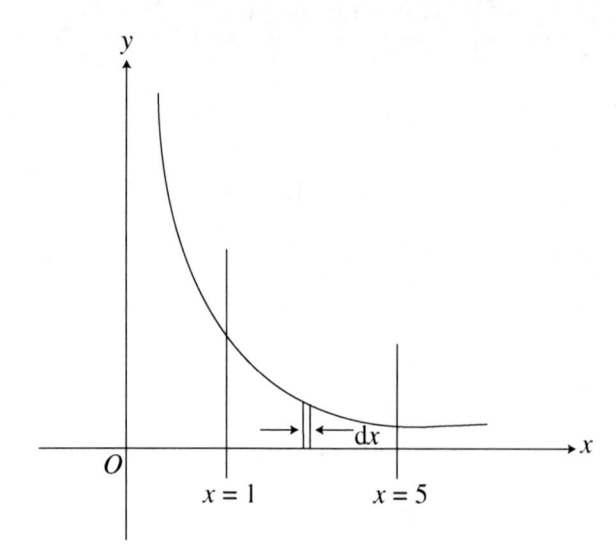

Consider an elemental strip of width dx and height y.

$$\pi \int_1^5 y^2 dx = \pi \int_1^5 \left(\frac{1}{x}\right)^2 dx$$

$$= \pi \int_1^5 x^{-2} dx$$

$$= \pi \left[\frac{x^{-1}}{-1}\right]_1^5 = \pi \left(-\frac{1}{5} + \frac{1}{1}\right)$$

$$= \pi \left[\frac{4}{5}\right] \text{ c.u.}$$

(b) $\int_0^3 \dfrac{x}{\sqrt{1+x^2}} dx = \int_1^{10} \dfrac{x}{u^{\frac{1}{2}}} \dfrac{du}{2x}$ where the limits change to

$u = 1 + 3^2 = 10$ and $u = 1 + 0^2 = 1$

let $u = 1 + x^2 = \dfrac{1}{2}\int_1^{10} \dfrac{du}{u^{\frac{1}{2}}} = \dfrac{1}{2}\int_1^{10} u^{-\frac{1}{2}} du$

$$= \frac{1}{2}\left[\frac{u^{\frac{1}{2}}}{\frac{1}{2}}\right]_1^{10} = \frac{1}{2}\left[2\sqrt{10} - 2\right]$$

$$= (\sqrt{10} - 1) \text{ s.u.}$$

Since $\dfrac{du}{dx} = 2x \Rightarrow dx = \dfrac{du}{2x}$.

3. (a) $\dfrac{dx}{dt} = 1 - \cos t \quad \dfrac{dy}{dt} = \sin t \Rightarrow \dfrac{dy}{dx} = \dfrac{\sin t}{1 - \cos t}.$

(b) $x = t - \sin t \Rightarrow \sin t = t - x \quad \cdots (1)$

$y = 1 - \cos t \Rightarrow \cos t = 1 - y \quad \cdots (2)$

Using the fundamental trigonometric identity $\sin^2 t + \cos^2 t \equiv 1$
squaring both equations of (1) and (2)

$\sin^2 t = (t - x)^2$ and $\cos^2 t = (1 - y)^2$

then $\sin^2 t + \cos^2 t = (t - x)^2 + (1 - y)^2 = 1$

$(t - x)^2 = 1 - (1 - y)^2$

$= 1 - 1 + 2y - y^2 = 2y - y^2$

$t - x = \pm\sqrt{y(2 - y)}$

$t = x \pm \sqrt{y(2 - y)}$

$\cos t = \cos\left[x \pm \sqrt{y(2 - y)}\right] =$

$1 - y = \cos\left[x \pm \sqrt{y(2 - y)}\right]$

$1 - \cos\left[x \pm \sqrt{y(2 - y)}\right] = y$

$0 \leq y \leq 2$

and $0 \leq x \leq 2\pi$.

---

4. (a) $(x + y)^{\frac{1}{2}} = x^{\frac{1}{2}}(1 + \dfrac{y}{x})^{\frac{1}{2}}$

$= \sqrt{x}\left(1 + \dfrac{1}{2}\dfrac{y}{x} + \dfrac{1}{2}\left(-\dfrac{1}{2}\right)\left(\dfrac{y}{x}\right)^2 \dfrac{1}{2!}\right)$

$\approx \sqrt{x}\left(1 + \dfrac{y}{2x}\right)$

$\approx \sqrt{x} + \dfrac{y}{2\sqrt{x}}$

and since y is small compared with x then $(\dfrac{y}{x})^2 << 1$

$\therefore (x + y)^{\frac{1}{2}} = -\sqrt{x} \approx \dfrac{y}{2}x^{-\frac{1}{2}}.$

(b) $(27.003)^{\frac{1}{3}} = (27 + 0.003)^{\frac{1}{3}} = \left[27\left(1 + \frac{0.003}{27}\right)\right]^{\frac{1}{3}}$

$= 27^{\frac{1}{3}}\left(1 + \frac{0.003}{27}\right)^{\frac{1}{3}}$

$= 3\left[1 + \frac{1}{3}\frac{0.001}{9} + \frac{1}{3}\left(-\frac{2}{3}\right)\left(\frac{0.001}{9}\right)^2\frac{1}{2}\right]$

$= 3\left(1 + \frac{0.001}{27} - \frac{1}{729}0.000001\right)$ neglecting the square term

$\approx 3 + \frac{0.001}{9} = 3.0001$.

---

5. (a)

x	-2	-1	0	1	2
$\dfrac{e^x}{1+e^x}$	0.119	0.269	0.5	0.731	0.881

$\int_{-2}^{2} \dfrac{e^x}{1+e^x}\,dx \approx \dfrac{h}{2}[y_1 + y_5 + 2(y_2 + y_3 + y_4)]$

$= \dfrac{1}{2}[0.119 + 0.881 + 2(0.269 + 0.5 + 0.731)]$

$= \dfrac{1}{2}(1 + 3) = 2$ square units.

(b) $\int_{-2}^{2} \dfrac{e^x}{1+e^x}\,dx = \int \dfrac{e^x}{u}\dfrac{du}{e^x} = \int \dfrac{du}{u} = \ln u$

$= \left[\ln(1 + e^x)\right]_{-2}^{2}$

$= \ln(1 + e^2) - \ln(1 + e^{-2})$

$= \ln\dfrac{1 + e^2}{1 + e^{-2}} = \ln\dfrac{8.389056099}{1.135335283}$

$= 2$ s.u.

since $u = 1 + e^x \Rightarrow e^x = u - 1$

$\dfrac{du}{dx} = e^x$.

6. (a)
$$\mathbf{r} = \begin{pmatrix} 2 \\ 3 \\ 5 \end{pmatrix} + \lambda \begin{pmatrix} 1 \\ 1 \\ 2 \end{pmatrix}$$

$$\mathbf{r} = \begin{pmatrix} 0 \\ 4 \\ 6 \end{pmatrix} + \mu \begin{pmatrix} -1 \\ 2 \\ 3 \end{pmatrix}$$

Equaling the coefficients of **i, j, k** for the two lines we have

$$2 + \lambda = -\mu \quad \Rightarrow \quad \lambda + \mu = -2 \quad \ldots (1)$$
$$3 + \lambda = 4 + 2\mu \quad \Rightarrow \quad \lambda - 2\mu = 1 \quad \ldots (2)$$
$$5 + 2\lambda = 6 + 3\mu \quad \Rightarrow \quad 2\lambda - 3\mu = 1 \quad \ldots (3)$$
$$(1) - (2) \quad 3\mu = -3 \quad \Rightarrow \quad \mu = -1 \text{ and } \lambda = -1$$

Substituting these values in (3) $-2 + 3 = 1$ therefore the two vector equations are equal and the lines interect.

(b) $\mathbf{r}_1 = \mathbf{r}_2$ and the position vector of the point of intersection is

$$\mathbf{r}_1 = \begin{pmatrix} 1 \\ 2 \\ 3 \end{pmatrix} = \mathbf{r}_2 = \begin{pmatrix} 0+1 \\ 4-2 \\ 6-3 \end{pmatrix} = \begin{pmatrix} 1 \\ 2 \\ 3 \end{pmatrix}$$

$\therefore (1, 2, 3)$.

$(\mathbf{i} + \mathbf{j} + 2\mathbf{k}) \cdot (-\mathbf{i} + 2\mathbf{j} + 3\mathbf{k}) = \sqrt{1^2 + 1^2 + 2^2}\sqrt{(-1)^2 + 2^2 + 3^2} \cos\theta$

$\cos\theta = \dfrac{-1+2+6}{\sqrt{6}\sqrt{14}} = \dfrac{7}{\sqrt{84}} = \dfrac{7}{\sqrt{7 \times 12}}$

$= \dfrac{7}{\sqrt{7}\sqrt{12}} \times \dfrac{\sqrt{7}\sqrt{12}}{\sqrt{7}\sqrt{12}} = \dfrac{\sqrt{84}}{12} = \sqrt{\dfrac{84}{144}} = \sqrt{\dfrac{4 \times 3 \times 7}{12 \times 12}} = \sqrt{\dfrac{7}{12}}.$

7.
$$A = \pi r^2$$

differentiating A with respect to r, we have $\quad \dfrac{dA}{dr} = 2\pi r$

dividing dA and dr by dt, we have $\quad \dfrac{\frac{dA}{dt}}{\frac{dr}{dt}} = 2\pi r$

the rate of change of $A = 2\pi r \times$ the rate of change of r.

TOTAL FOR PAPER: 75 MARKS

GCE Examinations

Test Paper 4

Advanced Level

Core Mathematics C4

Time: 1 hour 30 minutes

Instructions and Information

Candidates may use any calculator allowed by the regulations of their Examination Board.

Full marks are awarded for correct answers to ALL questions.

This paper has seven questions.

You can start working with any question and you must label clearly all parts.

1. Given that
$$f(x) = \frac{3+7x}{(1+2x)(1+3x)} \equiv \frac{A}{1+2x} + \frac{B}{1+3x},$$
   (a) find the values of the constant A and B. (8)
   (b) Hence, or otherwise, find the series expression in ascending powers of x, up to and including the term in x^2, of $\frac{3+7x}{(1+2x)(1+3x)}$. (4)

---

2. a) Find the exact value of the integral
$$\int_0^{\frac{1}{2}} \frac{4}{\sqrt{1-4x^2}} \, dx$$
   use the substitution $2x = \sin\theta$. (6)

   b) Determine the definite integral
$$\int_0^2 \ln x \, dx.$$
   (4)

---

3. (a) Differentiate $y = \dfrac{10}{x}$

   (i) implicitly   (ii) explicitly

   hence determine $\dfrac{d^2y}{dx^2}$. (6)

   (b) Differentiate $y \tan x = x^2$

   (i) implicitly   (ii) explicitly. (8)

---

4. The period of a simple pendulum is given by the formula $T = 2\pi\sqrt{\dfrac{l}{g}}$ where l is the length of the pendulum and g is the acceleration due to gravity.

   It is required to calculate g from the formula above, if errors of $+1\%$ in T and -0.5% in l are made, use the binomial expansion to determine the percentage error in the calculated value of g, giving your answer correct to three decimal places. (10)

---

5. The cartesian equation of an ellipse is given as $4x^2 + 9y^2 = 36$.

   If x is increasing at the rate of 0.1 cm/s, find the rate of decrease of y when $x = 2$. (12)

6. The position vectors of three points A, B and C are :

i − j,   2i + 3k   and − 2j − 3k   respectively.

Prove that A, B and C lie on the same line. (12)

---

7. (a) Use the trapezium rule with 6 intervals to find an approximation to

$$\int_0^6 \frac{1}{\sqrt{1+x^2}}\, dx.$$ (7)

(b) By separating the variables solve the differential equation

$$\frac{dx}{dt} = \frac{\sin^5 2x}{\cos 2x}$$

when $t = 1$, $x = \frac{\pi}{4}$ show that

$$\boxed{t = \frac{9}{8} - \frac{1}{8}\operatorname{cosec}^4 2x}$$ (8)

TOTAL FOR PAPER: 75 MARKS

GCE Examinations

Test Paper 4 Solutions

Advanced Level

Core Mathematics C4

1. (a) $\dfrac{3+7x}{(1+2x)(1+3x)} \equiv \dfrac{A}{1+2x} + \dfrac{B}{1+3x}$

$\dfrac{3+7x}{(1+2x)(1+3x)} \equiv \dfrac{A(1+3x) + B(1+2x)}{(1+2x)(1+3x)}$

$3 + 7x \equiv A(1+3x) + B(1+2x).$

Using the cover up rule

if $x = -\dfrac{1}{3}$

$3 + 7\left(-\dfrac{1}{3}\right) = B\left[1 + 2\left(-\dfrac{1}{3}\right)\right]$

$3 - \dfrac{7}{3} = B\left(1 - \dfrac{2}{3}\right) = \dfrac{1}{3}B$

$\dfrac{9-7}{3} = \dfrac{1}{3}B \quad \boxed{B = 2}$

and if $x = -\dfrac{1}{2}$

$3 + 7\left(-\dfrac{1}{2}\right) = A\left[1 + 3\left(-\dfrac{1}{2}\right)\right]$

$3 - \dfrac{7}{2} = A\left(1 - \dfrac{3}{2}\right)$

$-\dfrac{1}{2} = A\left(-\dfrac{1}{2}\right) \quad \boxed{A = 1}$

$\therefore \dfrac{3+7x}{(1+2x)(1+3x)} \equiv \dfrac{1}{1+2x} + \dfrac{2}{1+3x}.$

(b) $\dfrac{1}{1+2x} + \dfrac{2}{1+3x} = (1+2x)^{-1} + 2(1+3x)^{-1}$

$= 1 + (-1)2x + \dfrac{(-1)(-2)}{2!}(2x)^2$

$\quad + 2\left[1 + (-1)3x + \dfrac{(-1)(-2)(3x)^2}{2!}\right]$

$= 1 - 2x + 4x^2 + 2 - 6x + 9x^2 \times 2$

$= 3 - 8x + 22x^2.$

2. (a) $\int_0^{\frac{1}{2}} \frac{4}{\sqrt{1-4x^2}} \, dx$

let $2x = \sin\theta$, when $x = 0, \theta = 0$; when $x = \frac{1}{2}, \theta = \frac{\pi}{2}$

$2\frac{dx}{d\theta} = \cos\theta \Rightarrow dx = \frac{\cos\theta \, d\theta}{2}$

$\int_0^{\frac{\pi}{2}} \frac{4}{\sqrt{1-\sin^2\theta}} \frac{\cos\theta \, d\theta}{2}$

$2\int_0^{\frac{\pi}{2}} \frac{\cos\theta}{\cos\theta} \, d\theta$

$\int_0^{\frac{\pi}{2}} 2 \, d\theta = [2\theta]_0^{\frac{\pi}{2}} = \pi$ square units.

(b) $\int_1^2 \ln x \, dx = [x \ln x]_1^2 - \int_1^2 x \cdot \frac{1}{x} \, dx$

$= 2\ln 2 - \ln 1 - [x]_1^2 = 2\ln 2 - 1.$

---

3. (a) $y = \frac{10}{x}$

(i) $xy = 10$ differentiating implicitly with respect to x

$\frac{dx}{dx}y + x\frac{dy}{dx} = 0$

$x\frac{dy}{dx} = -y$

$\frac{dy}{dx} = -\frac{y}{x} = -\frac{\frac{10}{x}}{x} = -\frac{10}{x^2} \quad \therefore \frac{dy}{dx} = -\frac{10}{x^2}.$

(ii) Differentiating explicitly $y = \frac{10}{x} = 10x^{-1}$

$\frac{dy}{dx} = -10x^{-2} = -\frac{10}{x^2} = -10x^{-2}$

$\frac{d^2y}{dx^2} = 20x^{-3} = \frac{20}{x^3}.$

(b) (i) $y \tan x = x^2$

differentiating with respect to x

$$\frac{dy}{dx} \tan x + y \sec^2 x \frac{dx}{dx} = 2x \frac{dx}{dx}$$

$$\frac{dy}{dx} \tan x + y \sec^2 x = 2x$$

$$\frac{dy}{dx} = \frac{2x - y \sec^2 x}{\tan x}$$

$$= \frac{2x - \frac{x^2 \sec^2 x}{\tan x}}{\tan x} = \frac{2x \tan x - x^2 \sec^2 x}{\tan^2 x}$$

(ii) $y \tan x = x^2 \quad y = \frac{x^2}{\tan x}$

$$\frac{dy}{dx} = \frac{2x \tan x - x^2 \sec^2 x}{\tan^2 x}$$

$$= \frac{2x \tan x - x^2(1 + \tan^2 x)}{\tan^2 x}.$$

---

4. $T = 2\pi \sqrt{\dfrac{l}{g}} \quad T^2 = 4\pi^2 \dfrac{l}{g} \Rightarrow g = \dfrac{kl}{T^2}$

where $k = 4\pi^2$

$$T' = T + \frac{1}{100}T = T(1 + 0.01)$$

$$l' = l - \frac{0.5}{100}l = l(1 - 0.005)$$

$$g' = \frac{kl'}{(T')^2} = kl'(T')^{-2}$$

$$= kl(1 - 0.005)(1 + 0.01)^{-2} T^{-2}$$

$$= klT^{-2}(1 - 0.005)(1 + 0.01)^{-2}$$

$$= g(1 - 0.005)(1 + 0.01)^{-2}$$
$$= g(1 - 0.005)(1 - 0.02)$$
$$= g(1 - 0.02 - 0.005 + 0.0001) = g(1 - 0.0249).$$

The error in g is -0.0249 or -2.5%.

5.
$$4x^2 + 9y^2 = 36 \quad \cdots (1)$$

differentiating with respect to t

$$8x\frac{dx}{dt} + 18y\frac{dy}{dt} = 0$$

$$\frac{dx}{dt} = 0.1 \text{ cm/s}$$

$$8(2)\,0.1 + 18y\frac{dy}{dt} = 0 \quad \cdots (2)$$

when $x = 2$, substituting in (1)

$$4(2)^2 + 9y^2 = 36$$

$$9y^2 = 36 - 16 = 20$$

$$y^2 = \frac{20}{9}$$

$$y = \pm\frac{\sqrt{20}}{3} = \pm 1.49 \text{ to 3 s.f.}$$

From (2) $\quad 1.6 + 18y\dfrac{dy}{dt} = 0$

$$18y\frac{dy}{dt} = -1.6$$

$$\frac{dy}{dt} = -\frac{1.6}{18y} = -\frac{1.6}{18(1.49)}$$

$$\frac{dy}{dt} = -0.059656972 = -0.0597 \text{ to 3 s.f.}$$

the rate of decrease is 0.0597.

6.

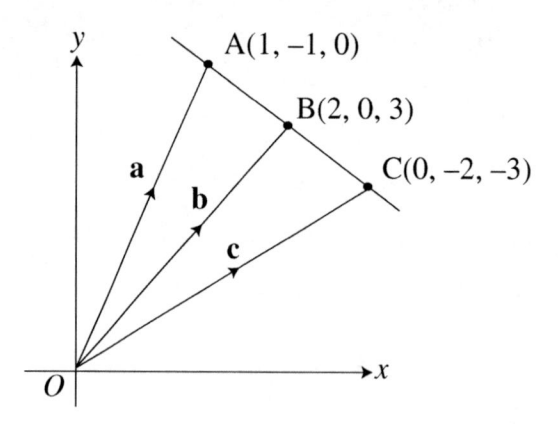

The position vectors of B and C are $\overrightarrow{OB} = 2\mathbf{i} - 3\mathbf{k}$, $\overrightarrow{OC} = -2\mathbf{i} - 3\mathbf{k}$ and the equation of the line $\overrightarrow{BC}$ is given $\mathbf{r} = \mathbf{b} + \lambda(\mathbf{c} - \mathbf{b})$

$$\mathbf{r} = \begin{pmatrix} 2 \\ 0 \\ 3 \end{pmatrix} + \lambda \begin{pmatrix} 0-2 \\ -2-0 \\ -3-3 \end{pmatrix} = \begin{pmatrix} 2-2\lambda \\ -2\lambda \\ 3-6\lambda \end{pmatrix}.$$

If A is on the line $\overrightarrow{BC}$ then

$$\mathbf{r} = \mathbf{i} - \mathbf{j} = \begin{pmatrix} 1 \\ -1 \\ 0 \end{pmatrix}$$

$\mathbf{i} - \mathbf{j} = (2 - 2\lambda)\mathbf{i} - 2\lambda\mathbf{j} + (3 - 6\lambda)\mathbf{k}$

$2 - 2\lambda = 1 \Rightarrow \lambda = \dfrac{1}{2}$

$-2\lambda = -1 \Rightarrow \lambda = \dfrac{1}{2}$

$3 - 6\lambda = 0 \Rightarrow \lambda = \dfrac{1}{2}.$

Therefore A, B and C are colinear.

---

7. (a) $\displaystyle\int_0^6 \dfrac{1}{\sqrt{1+x^2}}\, dx \approx \dfrac{h}{2}\left[y_1 + y_7 + 2(y_2 + y_3 + y_4 + y_5 + y_6)\right]$

$= \dfrac{1}{2}[1 + 0.068 + 2(0.707 + 0.333 + 0.189 + 0.124 + 0.089)]$

x	0	1	2	3	4	5	6
$\dfrac{1}{\sqrt{1+x^2}}$	1	0.707	0.333	0.189	0.124	0.089	0.068

$$\therefore \int_0^6 \frac{1}{\sqrt{1+x^2}}\, dx = \frac{1}{2}(1.068 + 2.884)$$

$$= 1.976 \text{ square units to 3 d.p.}$$

(b) $\dfrac{dx}{dt} = \dfrac{\sin^5 2x}{\cos 2x}$

$$\int \frac{\cos 2x}{\sin^5 2x}\, dx = \int dt = t$$

$$\frac{1}{2}\int \frac{d(\sin 2x)}{\sin^5 2x} = t + c$$

$$\frac{1}{2}\int (\sin 2x)^{-5}\, d(\sin 2x) = t + c$$

$$\frac{1}{2}\frac{(\sin 2x)^{-4}}{-4} = t + c$$

$$-\frac{1}{8\sin^4 2x} = t + c$$

$$-\frac{1}{8\sin^4 2\frac{\pi}{4}} = 1 + c$$

$$c = -\frac{1}{8} - 1 = -\frac{9}{8}$$

$$t = \frac{9}{8} - \frac{1}{8}\operatorname{cosec}^4 2x.$$

TOTAL FOR PAPER: 75 MARKS

GCE Examinations

Test Paper 5

Advanced Level

Core Mathematics C4

Time: 1 hour 30 minutes

Instructions and Information

Candidates may use any calculator allowed by the regulations of their Examination Board.

Full marks are awarded for correct answers to ALL questions.

This paper has seven questions.

You can start working with any question and you must label clearly all parts.

1. Express $\frac{3x^3+6x^2+6x+2}{(x+1)^3(2x+1)} = f(x)$ into partial fractions and

   show that $f(x) = \frac{1}{(x+1)^3} - \frac{1}{(x+1)^2} + \frac{1}{x+1} + \frac{1}{2x+1}$. (14)

2. (a) If $\frac{1}{x^2-3x+2} \equiv \frac{A}{x-2} + \frac{B}{x-1}$, find A and B, and hence (4)

   (b) Determine the indefinite integral $\int \frac{1}{x^2-3x+2} \, dx$. (4)

3. (a) The parametric equations of a curve are
   $$x = t^2 + 1$$
   $$y = t(t^2 + 1).$$
   Determine the cartesian equation. (6)

   (b) Show that $\frac{dy}{dx} = \frac{3x-2}{2\sqrt{x-1}}$ by either using the cartesian equation or the parametric equations. (6)

4. (a) Expand $(1 - 2x)^{-2}$, $|x| < \frac{1}{2}$, in ascending powers of x up to and including the term in x^3, simplifying each term. (4)

   (b) Hence find the first three terms in the expansion of $\frac{x+1}{(1-2x)^2}$. (4)

5. The value £C of a car t years after 1st January 2004 is given by the formula
   $C = 30000 \times (1.3)^{-t}$.

   Find the value of $\frac{dC}{dt}$ when $t = 5$. (6)

6. Three points A, B and C are collinear, the position vectors of A and C are given $\overrightarrow{OA} = 2\mathbf{i} - 3\mathbf{j} + 4\mathbf{k}$ and $\overrightarrow{OC} = -3\mathbf{i} + 2\mathbf{j} + 5\mathbf{k}$,
   find the position vector of B, for $\lambda = -1$. (8)

7. Evaluate:

   (a) $\int_2^5 \frac{x+3}{x-1} \, dx$. (6)

   (b) $\int_0^{2\pi} \sec^2 x \sin x \, dx$. (6)

   (c) $\int_0^2 (x+1) \ln \sqrt{x+1} \, dx$. (7)

TOTAL FOR PAPER: 75 MARKS

GCE Examinations

Test Paper 5 Solutions

Advanced Level

Core Mathematics C4

1. $\dfrac{3x^3 + 6x^2 + 6x + 2}{(x+1)^3(1+2x)} \equiv \dfrac{A}{(x+1)^3} + \dfrac{B}{(x+1)^2} + \dfrac{C}{x+1} + \dfrac{D}{2x+1}$

$3x^3 + 6x^2 + 6x + 2 \equiv A(2x+1) + B(x+1)(2x+1) + C(x+1)^2(2x+1) + D(x+1)^3$

by the cover up rule.

$$\text{if } x = -1$$
$$-3 + 6 - 6 + 2 = -A$$
$$-1 = -A$$
$$\boxed{A = 1}$$

$$\text{if } x = -\dfrac{1}{2}$$
$$-\dfrac{3}{8} + \dfrac{3}{2} - 3 + 2 = D\left(\dfrac{1}{8}\right)$$
$$\dfrac{1}{8} = \dfrac{D}{8}$$
$$\boxed{D = 1}$$

$$\text{if } x = 0$$
$$2 = A + B + C + 1 \Rightarrow A + B + C = 1$$
$$\text{since } A = 1, B + C = 0 \Rightarrow B = -C.$$

$$\text{if } x = 1$$
$$3 + 6 + 6 + 2 = 3A + 6B + 12C + 8D$$
$$17 = 3 - 6C + 12C + 8$$
$$6 = 6C$$
$$\boxed{C = 1}$$
$$\boxed{B = -1}$$

$\therefore \dfrac{3x^3 + 6x^2 + 6x + 2}{(x+1)^3(1+2x)} \equiv \dfrac{1}{(x+1)^3} - \dfrac{1}{(x+1)^2} + \dfrac{1}{x+1} + \dfrac{1}{2x+1}.$

2. (a)
$$\frac{1}{x^2 - 3x + 2} = \frac{1}{(x-2)(x-1)} \equiv \frac{A}{x-2} + \frac{B}{x-1}$$

$$1 \equiv A(x-1) + B(x-2)$$

by the cover up rule

if $x = 1$

$$1 = -B \Rightarrow \boxed{B = -1}$$

if $x = 2$

$$\boxed{1 = A}$$

(b) $\therefore \int \dfrac{1}{x^2 - 3x + 2}\, dx = \int \left(\dfrac{1}{x-2} - \dfrac{1}{x-1} \right) dx$

$$= \ln|x-2| - \ln|x-1| + \ln A$$

$$= \ln \frac{A(x-2)}{x-1}.$$

---

3 (a)
$$x = t^2 + 1 \quad \ldots (1)$$
$$y = t(t^2 + 1) \quad \ldots (2)$$

Eliminating t from the parametric equations.
Substitute (1) in (2)

$$y = t(x)$$
$$t = \frac{y}{x} \quad \ldots (3)$$

Substitute (3) in (1)

$$x = \left(\frac{y}{x}\right)^2 + 1$$

$$x = \frac{y^2}{x^2} + 1 \Rightarrow x^3 = y^2 + x^2$$

$$y^2 = x^3 - x^2 = x^2(x-1)$$

$$\therefore \boxed{y^2 = x^2(x-1)} \quad \ldots (4)$$

(b) differentiating (4) with respect to x

$$2y\frac{dy}{dx} = 2x(x-1) + x^2 \cdot 1$$

$$2y\frac{dy}{dx} = 2x^2 - 2x + x^2 = 3x^2 - 2x$$

$$\frac{dy}{dx} = \frac{3x^2 - 2x}{2y} = \frac{3x^2 - 2x}{2\sqrt{x^2(x-1)}} = \frac{3x-2}{2\sqrt{x-1}}.$$

or

$$\frac{dx}{dt} = 2t \text{ and } \frac{dy}{dt} = 3t^2 + 1$$

$$\frac{\frac{dy}{dt}}{\frac{dx}{dt}} = \frac{3t^2 + 1}{2t}$$

$$\frac{dy}{dx} = \frac{3t^2 + 1}{2t}$$

From (1)

$$t = \sqrt{x-1}$$

$$\frac{dy}{dx} = \frac{3(x-1) + 1}{2\sqrt{x-1}}$$

$$= \frac{3x - 3 + 1}{2\sqrt{x-1}} = \frac{3x-2}{2\sqrt{x-1}}$$

$$\frac{dy}{dx} = \frac{3x-2}{2\sqrt{x-1}}.$$

---

4. (a) $(1-2x)^{-2} = 1 + (-2)(-2x) + (-2)(-3)(-2x)^2\frac{1}{2!} + (-2)(-3)(-4)(-2x)^3\frac{1}{3!}$

$= 1 + 4x + 12x^2 + 32x^3.$

(b) $\dfrac{x+1}{(1-2x)^2} = (x+1)(1-2x)^{-2}$

$= (x+1)(1 + 4x + 12x^2 + 32x^3)$

$= x + 4x^2 + 12x^3 + 1 + 4x + 12x^2 + 32x^3$

$= 1 + 5x + 16x^2 + 44x^3.$

5. $\ln C = \ln 30000 - t \ln 1.3$

$$\frac{1}{C}\frac{dC}{dt} = -\ln 1.3$$

$$\frac{dC}{dt} = -C \ln 1.3 = -30000 \times (1.3)^{-t} \ln 1.3$$

where $t = 5$

$$\frac{dC}{dt} = -30000 \times (1.3)^{-5} \ln 1.3$$

$$= -2119.869735 = -2119.87 \text{ to 2 d.p.}$$

---

6. The vector equation of the line $\overrightarrow{AC}$ is given $\mathbf{r} = \mathbf{a} + \lambda(\mathbf{c} - \mathbf{a})$ since B lies on $\overrightarrow{AC}$ then $x_1\mathbf{i} + y_1\mathbf{j} + z_1\mathbf{k} = \mathbf{a} + \lambda(\mathbf{c} - \mathbf{a})$

$$= \lambda(\mathbf{c} - \mathbf{a}) = \lambda(-3\mathbf{i} + 2\mathbf{j} + 5\mathbf{k} - 2\mathbf{i} + 3\mathbf{j} - 4\mathbf{k}) = \lambda(-5\mathbf{i} + 5\mathbf{j} + \mathbf{k})$$

$$= \mathbf{a} + \lambda(\mathbf{c} - \mathbf{a}) = (2\mathbf{i} - 3\mathbf{j} + 4\mathbf{k}) + \lambda(-5\mathbf{i} + 5\mathbf{j} + \mathbf{k})$$

$$= (2 - 5\lambda)\mathbf{i} + (5\lambda - 3)\mathbf{j} + (4 + \lambda)\mathbf{k}$$

$x_1 = 2 - 5\lambda = 2 - 5(-1) = 7$

$y_1 = 5\lambda - 3 = 5(-1) - 3 = -8$

$z_1 = 4 + \lambda = 4 - 1 = 3$

therefore the position vector if B is

$$\boxed{\overrightarrow{OB} = 7\mathbf{i} - 8\mathbf{j} + 3\mathbf{k}}$$

---

7. (a) $\displaystyle\int_2^5 \frac{x+3}{x-1} dx = \int_2^5 \frac{x-1+4}{x-1} dx$

$$= \int_2^5 \left(1 + \frac{4}{x-1}\right) dx$$

$$= \left[x + 4\ln|x-1|\right]_2^5$$

$$= 5 + 4\ln 4 - (2 + 4\ln 1)$$

$$= 3 + 4\ln 4.$$

(b) $\displaystyle\int_0^{2\pi} \sec^2 x \sin x \, dx = \int_0^{2\pi} \frac{\sin x}{\cos^2 x} \, dx$

$\displaystyle = -\int_0^{2\pi} \frac{d(\cos x)}{\cos^2 x}$

$\displaystyle = -\int_0^{2\pi} (\cos x)^{-2} \, d(\cos x)$

$\displaystyle = \left[-\frac{(\cos x)^{-1}}{-1} \right]_0^{2\pi}$

$\displaystyle = \left[\frac{1}{\cos x} \right]_0^{2\pi} = \frac{1}{\cos 2\pi} - \frac{1}{\cos 0} = 1 - 1 = 0.$

(c) $\displaystyle\int_0^2 \underset{①}{(x+1)} \underset{②}{\ln \sqrt{x+1}} \, dx = \left[\frac{1}{2}(x+1)^2 \ln \sqrt{x+1} \right]_0^2$

$\displaystyle \qquad - \int_0^2 \frac{1}{2}(x+1)^2 \cdot \frac{1}{2(x+1)} \, dx$

$\displaystyle = \frac{1}{2} \cdot 9 \ln \sqrt{3} - 0 - \left[\frac{1}{4} \cdot \frac{(x+1)^2}{2} \right]_0^2 = \frac{9}{2} \ln \sqrt{3} - \frac{9}{8} + \frac{1}{8}$

$\displaystyle = \frac{9}{2} \ln \sqrt{3} - 1.$

TOTAL FOR PAPER: 75 MARKS

GCE Examinations

Test Paper 6

Advanced Level

Core Mathematics C4

Time: 1 hour 30 minutes

Instructions and Information

Candidates may use any calculator allowed by the regulations of their Examination Board.

Full marks are awarded for correct answers to ALL questions.

This paper has seven questions.

You can start working with any question and you must label clearly all parts.

1. (a) Decompose $\dfrac{6x^3 - 7x^2 - 7x + 7}{(x-1)(2x-3)}$ into partial fractions. (8)

   (b) Decompose $\dfrac{1}{x^2 - 11x + 24}$ into partial fractions. (6)

2. (i) $\displaystyle\int \tan x \sec^2 x\, dx$ (4)

   (ii) $\displaystyle\int \dfrac{\cos x}{\sin x}\, dx$ (4)

   (iii) $\displaystyle\int \cos^3 x \sin x\, dx.$ (4)

3. A curve has parametric equations
$$x = 3\cos t$$
$$y = 3\cos 2t.$$

   (a) Determine the cartesian equation of the curve. (3)

   (b) Determine

   (i) $\dfrac{dy}{dx}$ and (4)

   (ii) $\dfrac{d^2y}{dx^2}.$

   (c) Use the parametric equations to find $\dfrac{dy}{dx}$ and $\dfrac{d^2y}{dx^2}$ and show that $\dfrac{d^2y}{dx^2} = \dfrac{4}{9}.$ (6)

4. The binomial expansion $(1 + 5x)^{-\frac{3}{4}}$

   in ascending powers of x up to and including the term in x^3 is
   $$1 + ax + bx^2 + cx^3, \quad |5x| < 1.$$

   (a) Find the values of a, b and c. (5)

   (b) Determine $(1.5)^{-\frac{3}{4}}$ giving your answer to 3 decimal places, and work out the percentage error. (5)

5. The radius of an oil drop is modelled by the equation

   $r = 2(1 - e^{-kt})$,   where k is a positive constant and $t > 0$.

   Show that $\dfrac{dA}{dt} = 8\pi(1 - e^{-kt})e^{-kt}$ (8)

   where $A = \pi r^2$ and r is in cm.

   Hence calculate the rate of the area A when $t = 2$ and $k = 1$ to three significant figures. (4)

---

6. The following vectors are given (i) $\mathbf{a} = 3\mathbf{i} + 5\mathbf{j} - \mathbf{k}$ (ii) $\mathbf{b} = -\mathbf{i} - 2\mathbf{j} + 3\mathbf{k}$.

   Determine the magnitudes of the vectors and hence find the corresponding unit vectors. (8)

---

7. Determine the volume generated by rotating the curve $y = \sin 2x$ between $x = 0$ and $x = \dfrac{\pi}{2}$. (6)

TOTAL FOR PAPER: 75 MARKS

GCE Examinations

Test Paper 6 Solutions

Advanced Level

Core Mathematics C4

1. (a) $\dfrac{6x^3 - 7x^2 - 7x + 7}{(x-1)(2x-3)}$

$$\dfrac{6x^3 - 7x^2 - 7x + 7}{2x^2 - 3x - 2x + 3} = \dfrac{6x^3 - 7x^2 - 7x + 7}{2x^2 - 5x + 3}$$

$$\begin{array}{r}
3x + 4 \\
2x^2 - 5x + 3 \overline{\smash{\big)}\, 6x^3 - 7x^2 - 7x + 7} \\
\underline{6x^3 - 15x^2 + 9x } \\
8x^2 - 16x + 7 \\
\underline{8x^2 - 20x + 12} \\
4x - 5
\end{array}$$

$\therefore \dfrac{6x^3 - 7x^2 - 7x + 7}{(x-1)(2x-3)} = 3x + 4 + \dfrac{4x - 5}{(x-1)(2x-3)}$

$\dfrac{4x - 5}{(x-1)(2x-3)} \equiv \dfrac{A}{x-1} + \dfrac{B}{2x-3}$

$4x - 5 \equiv A(2x - 3) + B(x - 1)$

using the cover up rule

if $x = 1$

$$4 - 5 = A(-1)$$

$$-1 = -A$$

$$\boxed{A = 1}$$

if $x = \tfrac{3}{2}$

$$4\left(\dfrac{3}{2}\right) - 5 = B\left(\dfrac{1}{2}\right)$$

$$1 = \dfrac{B}{2}$$

$$\boxed{B = 2}$$

$\dfrac{6x^3 - 7x^2 - 7x + 7}{(x-1)(2x-3)} \equiv 3x + 4 + \dfrac{1}{x-1} + \dfrac{2}{2x-3}$

(b) $\dfrac{1}{x^2 - 11x + 24} = \dfrac{1}{(x-3)(x-8)} \equiv \dfrac{A}{x-3} + \dfrac{B}{x-8}$

$$1 \equiv A(x-8) + B(x-3)$$

if $x = 8$ \qquad\qquad if $x = 3$

$1 = 5B$ \qquad\qquad $1 = A(-5)$

$B = \dfrac{1}{5}$ \qquad\qquad $A = -\dfrac{1}{5}$

$$\dfrac{1}{x^2 - 11x + 24} \equiv -\dfrac{1}{5(x-3)} + \dfrac{1}{5(x-8)}.$$

---

2. (i) $\displaystyle\int \tan x \sec^2 x \, dx$

$= \displaystyle\int \tan x \, d(\tan x)$

$= \dfrac{\tan^2 x}{2} + c$

(ii) $\displaystyle\int \dfrac{\cos x}{\sin x} dx = \int \dfrac{d(\sin x)}{\sin x} = \ln \sin x + c$

(iii) $\displaystyle\int \cos^3 x \sin x \, dx = -\int \cos^3 x \, d(\cos x)$

$= -\dfrac{\cos^4 x}{4} + c.$

---

3. (a) $x = 3\cos t$

$y = \cos 2t = 2\cos^2 t - 1 = 2\left(\dfrac{x}{3}\right)^2 - 1$

$y = \dfrac{2x^2}{9} - 1.$

(b) (i) $\dfrac{dy}{dx} = \dfrac{4}{9}x$ \qquad (ii) $\dfrac{d^2y}{dx^2} = \dfrac{4}{9}.$

(c) $\dfrac{dx}{dt} = -3\sin t \qquad \dfrac{dy}{dt} = -2\sin 2t$

$\dfrac{dy}{dx} = \dfrac{2\sin 2t}{3\sin t} = \dfrac{2}{3} \times \dfrac{2\sin t \cos t}{\sin t}$

$= \dfrac{4}{3}\cos t = \dfrac{4}{3}\dfrac{x}{3} = \dfrac{4}{9}x$

$\dfrac{d^2y}{dx^2} = -\dfrac{4}{3}\sin t \dfrac{dt}{dx} = -\dfrac{4}{3} = \dfrac{\frac{dx}{dt}}{-3}\dfrac{dt}{dx} = \dfrac{4}{9}.$

---

4. (a) $(1+5x)^{-\frac{3}{4}} = 1 + \left(-\dfrac{3}{4}\right)(5x) + \left(-\dfrac{3}{4}\right)\left(-\dfrac{7}{4}\right)(5x)^2\dfrac{1}{2!}$

$+ \left(-\dfrac{3}{4}\right)\left(-\dfrac{7}{4}\right)\left(-\dfrac{11}{4}\right)(5x)^3\dfrac{1}{3!}$

$= 1 - \dfrac{15}{4}x + \dfrac{21}{16} \times \dfrac{25}{2}x^2 - \dfrac{231}{64} \times \dfrac{125}{6}x^3$

$= 1 - 3.75x + 16.4025x^2 - 75.1953125x^3.$

$\therefore a = -3.75,\ b = 16.4025 \text{ and } c = 75.1953125$

(b) $(1 + 5 \times 0.1)^{-\frac{3}{4}} = (1.5)^{-0.75}$

$= 1 - 0.375 + 0.1640625 - 0.0751953125$

$= 0.713867188$

$= 0.714 \text{ to 3 d.p.}$

The actual value of $(1.5)^{-0.75} = 0.737787946 = 0.738$ to 3 d.p.

In 0.737787946 the error is $0.737787946 - 0.713867187$

$= 0.023920758$, in 100 the error is 3.24 to 3 d.p.

$\therefore 3.24\%$ is the error.

5.
$$A = \pi r^2$$
$$\frac{dA}{dr} = 2\pi r$$
$$\frac{\frac{dA}{dt}}{\frac{dr}{dt}} = 2\pi r$$
$$\frac{dA}{dt} = 2\pi r \frac{dr}{dt}$$

$r = 2(1 - e^{-kt}) \Rightarrow \dfrac{dr}{dt} = 2ke^{-kt}$

$$\therefore \frac{dA}{dt} = 2\pi r \, 2ke^{-kt}$$
$$= 4\pi r e^{-kt}$$
$$= 4\pi 2(1 - e^{-kt}) e^{-kt}$$
$$= 8\pi (1 - e^{-kt}) e^{-kt}$$
$$\frac{dA}{dt} = 8\pi (1 - e^{-2}) e^{-2}$$
$$= 3.401346653 - 0.460322212 = 2.94102444$$
$$= 2.94 \text{ cm}^2/\text{s. to 3 s.f.}$$

6. (i) $\mathbf{a} = 3\mathbf{i} + 5\mathbf{j} - \mathbf{k}$

$|\mathbf{a}|$ denotes the magnitude or modulus of the vector $\mathbf{a}$

$|\mathbf{a}| = \sqrt{(3)^2 + (5)^2 + (-1)^2} = \sqrt{9 + 25 + 1} = \sqrt{35}$

$\hat{\mathbf{a}}$ denotes the unit vector in the direction of $\mathbf{a}$

$\hat{\mathbf{a}} = \dfrac{\mathbf{a}}{|\mathbf{a}|} = \dfrac{1}{\sqrt{35}}(3\mathbf{i} + 5\mathbf{j} - \mathbf{k})$

(ii) $\mathbf{b} = -\mathbf{i} - 2\mathbf{j} + 3\mathbf{k}$

$|\mathbf{b}| = \sqrt{(-1)^2 + (-2)^2 + 3^2} = \sqrt{14}$

$\hat{\mathbf{b}} = \dfrac{\mathbf{b}}{|\mathbf{b}|} = \dfrac{-\mathbf{i} - 2\mathbf{j} + 3\mathbf{k}}{\sqrt{14}}$

the unit vector in the direction of the vector $\mathbf{b}$.

7. $\pi \int_0^{\frac{\pi}{2}} \sin^2 x \, dx = \pi \int_0^{\frac{\pi}{2}} \frac{1 - \cos 4x}{2} \, dx = \frac{\pi}{2} \left[x - \frac{\sin 4x}{4} \right]_0^{\frac{\pi}{2}} = \frac{\pi}{2} \frac{\pi}{2} = \frac{\pi^2}{4}$

where $\cos 4x = 1 - 2\sin^2 2x \Rightarrow 2\sin^2 2x = 1 - \cos 4x$

$$\sin^2 2x = \frac{1 - \cos 4x}{2}$$

TOTAL FOR PAPER: 75 MARKS

GCE Examinations

Test Paper 7

Advanced Level

Core Mathematics C4

Time: 1 hour 30 minutes

Instructions and Information

Candidates may use any calculator allowed by the regulations of their Examination Board.

Full marks are awarded for correct answers to ALL questions.

This paper has seven questions.

You can start working with any question and you must label clearly all parts.

1. If $\dfrac{11x^2 - x + 4}{(x^2 + 1)(2x - 1)} \equiv \dfrac{Ax + B}{x^2 + 1} + \dfrac{C}{2x - 1}$

   determine the values of A, B and C (8)

   and check the result. (2)

---

2. (i) $\displaystyle\int x^2 \cos 2x \, dx$ (6)

   (ii) $\displaystyle\int x \ln x \, dx$. (4)

---

3. The parametric equations of a curve are
$$x = t^3 - 8t \text{ and } y = t^2.$$

   (a) Determine $\dfrac{dy}{dx}$. (4)

   (b) Sketch the curve. (4)

   (c) Determine the cartesian equation. (2)

---

4. Express (i) $(1 + x)^{\frac{1}{2}}$ and (ii) $(1 - x)^{-3}$ as series of ascending powers of x in each case up to and including the term in x^2. (6)

   Hence show that, if x is small $\dfrac{(1+x)^{\frac{1}{2}}}{(1-x)^3} = 1 + \dfrac{7}{2}x + \dfrac{59}{8}x^2$. (4)

   Calculate the percentage change which occurs in the value of $\dfrac{W^{\frac{1}{2}}}{Z^3}$,

   if W is increased by 1% and Z is decreased by 1%. (6)

---

5. The volume of an expanding sphere is increasing at the rate 100 mm³/s.

   Determine the rate at which the radius of the sphere is increasing at the instant when the radius is 75 mm. (6)

---

6. The position vectors of A and B are $\mathbf{a} = 2\mathbf{i} + 3\mathbf{j} - 4\mathbf{k}$ and $\mathbf{b} = -3\mathbf{i} - 2\mathbf{j} + \mathbf{k}$,

   find the position vector of the mid-point of AB.

   Determine the magnitude of $\overrightarrow{AB}$. (6)

Show that the position vector of point C dividing AB in the ratio $\lambda:\mu$ is given by $\overrightarrow{OC} = \dfrac{\mathbf{a}\mu + \mathbf{b}\lambda}{\lambda + \mu}$. (6)

7. Consider the function of $y = e^x$ from $x = 0$ to $x = 1$ as shown.

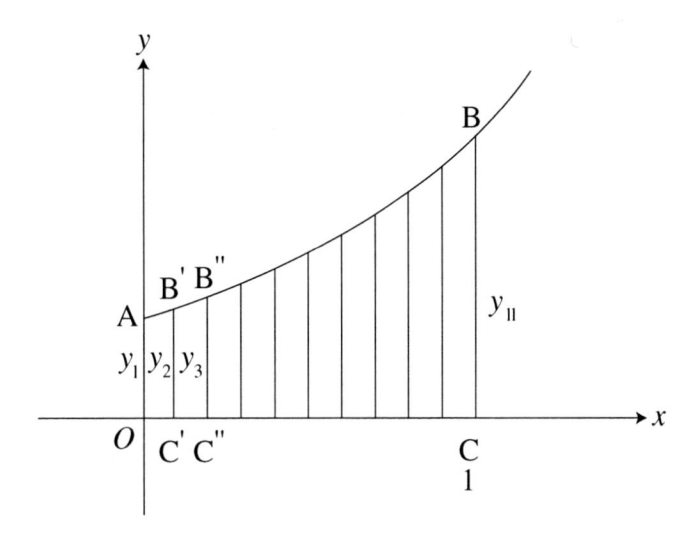

There are 11 ordinates or 10 equal intervals of width h.

The area of the trapezium OAB$'$C$'$ is given by the formula where OA = y_1, B$'$C$'$ = y_2 and OC$'$ = $h = 0.1$.

Area B$'$ B$''$ C$''$ C$'$ = $\dfrac{1}{2}(y_2 + y_3)h$ where B$''$ C$''$ is y_3.

Show that the Trapezium Rule is given

$$\int y\,dx = \dfrac{h}{2}[y_1 + y_{11} + 2(y_2 + y_3 + y_4 + y_5 + y_6 + y_7 + y_8 + y_9 + y_{10})]$$

hence determine $\displaystyle\int_0^1 e^x\,dx$.

TOTAL FOR PAPER: 75 MARKS

GCE Examinations

Test Paper 7 Solutions

Advanced Level

Core Mathematics C4

1. $\dfrac{11x^2 - x + 4}{(x^2 + 1)(2x - 1)} \equiv \dfrac{Ax + B}{x^2 + 1} + \dfrac{C}{2x - 1}$

$11x^2 - x + 4 \equiv (Ax + B)(2x - 1) + C(x^2 + 1)$

using the cover up rule

if $x = \dfrac{1}{2}$

$11\left(\dfrac{1}{2}\right)^2 - \left(\dfrac{1}{2}\right) + 4 = C\left(\dfrac{1}{4} + 1\right)$

$\dfrac{11}{4} - \dfrac{1}{2} + 4 = C\dfrac{5}{4}$

$25 = 5C$

$\boxed{C = 5}$

$11x^2 - x + 4 \equiv 2Ax^2 + 2Bx - Ax - B + 5x^2 + 5$

$\equiv (2A + 5)x^2 + (2B - A)x - B + 5.$

Equating coefficients

$11 = 2A + 5,$ $-1 = 2B - A,$ $4 = -B + 5$

$2A = 6$ $-1 = 2B - 3$ $-B = -1$

$\boxed{A = 3}$ $2 = 2B$ $\boxed{B = 1}$

$\boxed{B = 1}$

$\therefore \dfrac{11x^2 - x + 4}{(x^2 + 1)(2x - 1)} \equiv \dfrac{3x + 1}{x^2 + 1} + \dfrac{5}{2x - 1}$

$= \dfrac{(3x + 1)(2x - 1) + 5(x^2 + 1)}{(x^2 + 1)(2x - 1)}$

$= \dfrac{6x^2 + 2x - 3x - 1 + 5x^2 + 5}{(x^2 + 1)(2x - 1)}$

$= \dfrac{11x^2 - x + 4}{(x^2 + 1)(2x - 1)}.$

2. (i) $\int \underset{(2)}{x^2} \underset{(1)}{\cos 2x} \, dx = \frac{\sin 2x}{2} x^2 - \int \frac{\sin 2x}{2} 2x \, dx$

$$= \frac{1}{2}x^2 \sin 2x - \int \underset{(2)}{x} \underset{(1)}{\sin 2x} \, dx$$

$$= \frac{1}{2}x^2 \sin 2x - \left[-\frac{1}{2}x \cos 2x + \frac{1}{4}\sin 2x + c \right]$$

where

$\int \underset{(2)}{x} \underset{(1)}{\sin 2x} \, dx = \left(\frac{-\cos 2x}{2} \right) x - \int -\frac{\cos 2x}{2} .1 \, dx$

$$= -\frac{1}{2}x \cos 2x + \frac{1}{2} \int \cos 2x \, dx$$

$$= -\frac{1}{2}x \cos 2x + \frac{1}{2} \frac{\sin 2x}{2} + c$$

$\int x^2 \cos 2x = \frac{1}{2}x^2 \sin 2x + \frac{1}{2}x \cos 2x - \frac{1}{4}\sin 2x + k.$

(ii) $\int \underset{(1)}{x} \underset{(2)}{\ln x} \, dx = \frac{x^2}{2}\ln x - \int \frac{x^2}{2}.\frac{1}{x} \, dx$

$$= \frac{x^2}{2}\ln x - \frac{1}{2}\frac{x^2}{2} + c$$

$$= \frac{x^2 \ln x}{2} - \frac{1}{4}x^2 + c.$$

---

3. (a) $x = t^3 - 8t \quad \cdots (1)$

$y = t^2 \quad \cdots (2)$

Differentiating (1) and (2) with respect of t, we have

$\frac{dx}{dt} = 3t^2 - 8 \quad \cdots (3)$

$\frac{dy}{dt} = 2t \quad \cdots (4)$

Dividing (4) by (3)

$\frac{dy}{dx} = \frac{2t}{3t^2 - 8}.$

(b) When $t = 0$, $\quad x = 0$ and $y = 0$

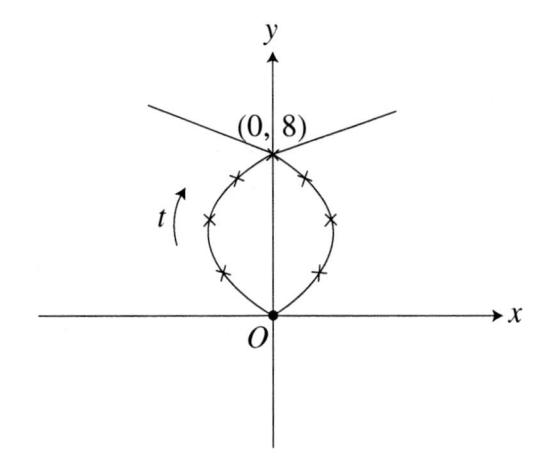

when $t = 1$, $\quad x = -7$ and $y = 1$

when $t = -1$, $\quad x = 7$ and $y = 1$

when $t = 2$, $\quad x = -8$ and $y = 4$

when $t = -2$, $\quad x = 8$ and $y = 4$

(c) $x = (\sqrt{y})^3 - 8(\sqrt{y}) \quad t = \sqrt{y}$

$x = y\sqrt{y} - 8\sqrt{y} = \sqrt{y}(y - 8)$.

---

4. (i) $(1+x)^{\frac{1}{2}} = 1 + \frac{1}{2}x + \frac{1}{2}\left(-\frac{1}{2}\right)x^2 \frac{1}{2!}$

$\qquad = 1 + \frac{1}{2}x - \frac{1}{8}x^2$

(ii) $(1-x)^{-3} = 1 + (-3)(-x) + (-3)(-4)(-x)^2 \frac{1}{2!}$

$\qquad = 1 + 3x + 6x^2$

$\dfrac{(1+x)^{\frac{1}{2}}}{(1-x)^3} = (1+x)^{\frac{1}{2}}(1-x)^{-3}$

$\qquad = \left(1 + \frac{1}{2}x - \frac{1}{8}x^2\right)(1 + 3x + 6x^2)$

$\qquad = 1 + 3x + 6x^2 + \frac{1}{2}x + \frac{3}{2}x^2 - \frac{1}{8}x^2 = 1 + \frac{7}{2}x + \frac{59}{8}x^2$.

$$\frac{W^{\frac{1}{2}}\left(1+\frac{1}{100}\right)^{\frac{1}{2}}}{Z^3\left(1-\frac{1}{100}\right)^3} = \frac{W^{\frac{1}{2}}}{Z^3}(1.01)^{\frac{1}{2}}(0.99)^{-3}$$

$$= \frac{W^{\frac{1}{2}}}{Z^3}\left(1 + \frac{7}{2} \times \frac{1}{100} + \frac{59}{2} \times \frac{1}{100^2}\right)$$

$$= \frac{W^{\frac{1}{2}}}{Z^3}(1 + 0.035 + 0.00295) = 1.03795 \frac{W^{\frac{1}{2}}}{Z^3}$$

3.8% the percentage change in $\frac{W^{\frac{1}{2}}}{Z^3}$ is 3.8 to 2 s.f.

---

5. $V = \frac{4}{3}\pi r^3 \qquad \frac{dV}{dr} = 4\pi r^2 = S \qquad \frac{dS}{dr} = 8\pi r$

$\frac{\frac{dV}{dt}}{\frac{dr}{dt}} = 4\pi r^2 = \frac{100}{\frac{dr}{dt}} \qquad \frac{dr}{dt} = \frac{1.0}{4\pi 75^2} = 1.42 \times 10^{-3}$ mm/s.

6.

A(2, 3, −4)

M

a

B(−3, −2, 1)

b

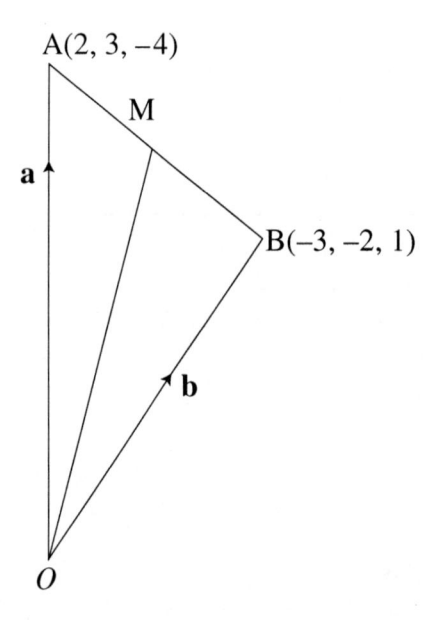

O

The midpoint M $\left(\frac{2-3}{2}, \frac{3-2}{2}, \frac{-4+1}{2}\right) \equiv M\left(-\frac{1}{2}, \frac{1}{2}, -\frac{3}{2}\right)$

hence the position vector of the mid-point is $\overrightarrow{OM} = -\frac{1}{2}\mathbf{i} + \frac{1}{2}\mathbf{j} - \frac{3}{2}\mathbf{k}$

$\vec{AB} = \mathbf{b} - \mathbf{a} = -3\mathbf{i} - 2\mathbf{j} + \mathbf{k} - 2\mathbf{i} - 3\mathbf{j} + 4\mathbf{k} = -5\mathbf{i} - 5\mathbf{j} + 5\mathbf{k}$

$|\vec{AB}| = \sqrt{(-5)^2 + (-5)^2 + 5^2} = \sqrt{75} = 5\sqrt{3}.$

$\dfrac{\vec{AC}}{\vec{CB}} = \dfrac{\lambda}{\mu}$

$\dfrac{\vec{AC}}{\vec{AB}} = \dfrac{\lambda}{\lambda + \mu}$

$\vec{AC} = \dfrac{\lambda}{\lambda + \mu} \vec{AB}$

$\vec{OC} = \vec{OA} + \vec{AC} = \mathbf{a} + \dfrac{\lambda}{\lambda + \mu}(\mathbf{b} - \mathbf{a})$

$= \dfrac{\mathbf{a}\lambda + \mathbf{a}\mu + \lambda\mathbf{b} - \lambda\mathbf{a}}{\lambda + \mu}$

$\vec{OC} = \dfrac{\mathbf{a}\mu + \mathbf{b}\lambda}{\lambda + \mu}.$

7. Area $OABC = AB'C'O + B'B''C''C' + \cdots$

$= \dfrac{h(y_1 + y_2)}{2} + \dfrac{h(y_2 + y_3)}{2} + \cdots + \dfrac{h(y_{10} + y_{11})}{2}$

$= \dfrac{h}{2}[y_1 + y_{11} + 2(y_2 + y_3 + y_4 + y_5 + \cdots + y_{10})]$

$\int e^x \, dx \approx \dfrac{0.1}{2}[e^0 + e^1 + 2(e^{0.1} + e^{0.2} + \cdots + e^{0.9})]$

$= 0.05[1 + 2.718281828 + 2(1.105170918 + 1.221402758 + 1.349858808$

$+ 1.491824698 + 1.648721271 + 1.822188 + 2.013752707$

$+ 2.225540478 + 2.459603111)]$

$= 0.05[1 + 2.718281828 + 30.675988] = 0.05 \times 3439426983 = 1.719713491$

$= 1.72$ to 3 s.f.

check $\displaystyle\int_0^1 e^x \, dx = [e^x]_0^1 = e - 1 = 1.72$ to s.f.

TOTAL FOR PAPER: 75 MARKS

GCE Examinations

Test Paper 8

Advanced Level

Core Mathematics C4

Time: 1 hour 30 minutes

Instructions and Information

Candidates may use any calculator allowed by the regulations of their Examination Board.

Full marks are awarded for correct answers to ALL questions.

This paper has seven questions.

You can start working with any question and you must label clearly all parts.

1. $f(x) = \dfrac{5x+2}{(1-3x)^2}$, $|x| < \dfrac{1}{3}$.

   Given that, for $x \neq \dfrac{1}{3}$,

   $\dfrac{5x+2}{(1-3x)^2} \equiv \dfrac{A}{(1-3x)^2} + \dfrac{B}{(1-3x)}$, where A and B are constants,

   (a) find the values of A and B. (5)

   (b) Hence or otherwise, find the series expansion of $f(x)$, in ascending powers of x, up to and including the term in x^3, simplifying each term. (7)

2.

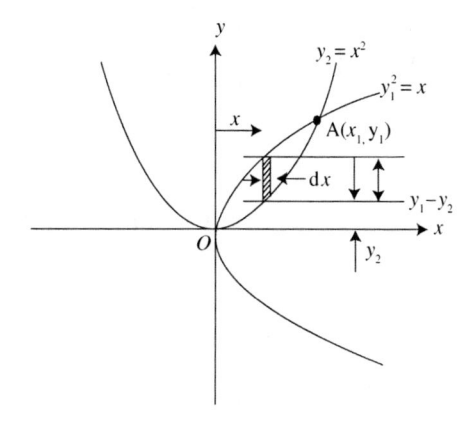

The sketch shows two parabolas intersecting at the origin O and A write down the coordinates of these two points.

By considering elemental strips y_1 and y_2 determine the area enclosed by the two curves and show that it is equal to $\dfrac{1}{3}$ square units (8)

3. (a) Sketch the curve represented by the parametric equations.

   $$x = 3t^2 \text{ and } y = 3t - 3t^3.$$ (6)

   (b) Determine $\dfrac{dx}{dt}$ and $\dfrac{dy}{dt}$ and hence $\dfrac{dy}{dx}$.

   What is the purpose of the parameters? (4)

   (c) By eliminating the parameter t from the x and y coordinates, obtain a cartesian equation for the curve, comment on the type of curve. (4)

4. $f(x) = \dfrac{1}{\sqrt{1+x}} + \sqrt{1-x} \quad -1 < x < 1.$

Find the series expansion of f(x), in ascending powers of x, up to and including the term in x^3. (6)

---

5. The surface area S and volume V of a solid sphere are changing with respect to t when it is uniformly heated.

When its surface area is increasing at a rate of 1 mm²/s, determine the rate at which the volume is increasing when its radius is 10 cm. (8)

---

6. if
$$\mathbf{a} = \begin{pmatrix} 1 \\ 2 \\ 3 \end{pmatrix}, \mathbf{b} = \begin{pmatrix} 2 \\ 2 \\ 2 \end{pmatrix}, \mathbf{c} = \begin{pmatrix} 3 \\ 0 \\ 5 \end{pmatrix}$$

write down the position vectors in terms of **i, j** and **k** for **a, b, c** and hence find

(i) $|\mathbf{a}|$ (2)

(ii) $|\mathbf{b} - \mathbf{a}|$ (3)

(iii) $|2\mathbf{c} - \tfrac{1}{2}\mathbf{a}|$. (4)

---

7. (i) $\displaystyle\int x\, 3^x\, dx$ (5)

(ii) $\displaystyle\int \sin x\, e^x \cos x\, dx$ (10)

(iii) $\displaystyle\int x^2 \ln 2x\, dx$ (5)

TOTAL FOR PAPER: 75 MARKS

GCE Examinations

Test Paper 8 Solutions

Advanced Level

Core Mathematics C4

1. (a) $\dfrac{5x+2}{(1-3x)^2} \equiv \dfrac{A}{(1-3x)^2} + \dfrac{B}{(1-3x)} = f(x)$

$5x + 2 \equiv A + B(1-3x)$

if $x = \tfrac{1}{3}$, $A = \tfrac{5}{3} + 2 = \tfrac{11}{3} \Rightarrow \boxed{A = \dfrac{11}{3}}$

if $x = 0$, $2 = \tfrac{11}{3} + B$ $\quad \boxed{B = -\dfrac{5}{3}}$

$\dfrac{5x+2}{(1-3x)^2} \equiv \dfrac{11}{3(1-3x)^2} - \dfrac{5}{3(1-3x)}$

$f(x) = \dfrac{11}{3}(1-3x)^{-2} - \dfrac{5}{3}(1-3x)^{-1}$

$= \dfrac{11}{3}\left[1 + (-2)(-3x) + \dfrac{(-2)(-3)(-3x)^2}{2!} + \dfrac{(-2)(-3)(-4)(-3x)^3}{3!} \right]$

$\quad - \dfrac{5}{3}\left[1 + (-1)(-3x) + \dfrac{(-1)(-2)}{2!}(-3x)^2 + \dfrac{(-1)(-2)(-3)}{3!}(-3x)^3 \right]$

$= \dfrac{11}{3}\left[1 + 6x + 27x^2 + 108x^3 \right] - \dfrac{5}{3}\left[1 + 3x + 9x^2 + 27x^3 \right]$

$= 2 + 17x + 84x^2 + 351x^3.$

2.

$y = x^2 = \sqrt{x} \quad \Rightarrow x^4 = x \quad \Rightarrow x(x^3 - 1) = 0$

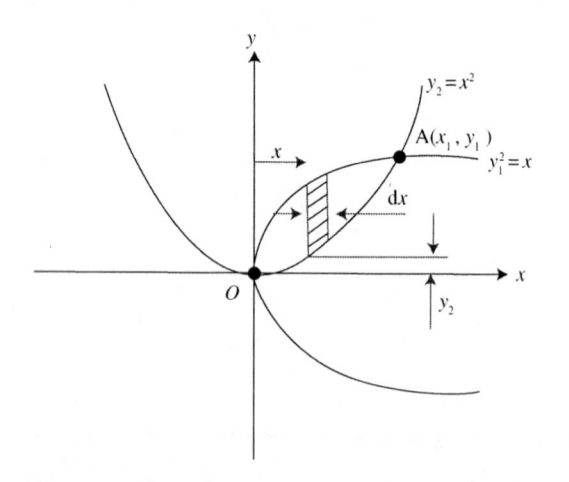

$x = 1$ and $y = 1 \quad \therefore A(1, 1)$

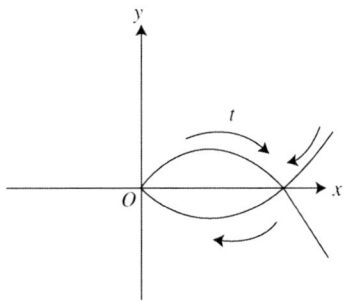

The area under the graph for $y_1^2 = x$ is given $\int_0^1 y_1 \, dx$ and that for $y_2 = x^2$ is $\int_0^1 y_2 \, dx$.

Area required $\int_0^1 y_1 \, dx - \int_0^1 y_2 \, dx$

$$= \int_0^1 \sqrt{x} - \int_0^1 x^2 \, dx$$

$$= \left[\frac{x^{\frac{3}{2}}}{\frac{3}{2}} - \frac{x^3}{3} \right]_0^1 = \frac{2}{3} - \frac{1}{3} = \frac{1}{3} \text{ s.u.}$$

---

3. When

$t = 0, \quad x = 0 \text{ and } y = 0$

$t = \dfrac{1}{4} \quad x = \dfrac{3}{16} \quad y = \dfrac{45}{64}$

$t = \dfrac{1}{2} \quad x = \dfrac{3}{4} \quad y = \dfrac{9}{8}$

$t = 1 \quad x = 3 \quad y = 0$

$t = 2 \quad x = 12 \quad y = -18$

$t = 3 \quad x = 27 \quad y = -72$

$t = -1 \quad x = 3 \quad y = 0$

$t = -2 \quad x = 12 \quad y = 18$

$t = -3 \quad x = 27 \quad y = 72$

$t = -\dfrac{1}{2} \quad x = \dfrac{3}{4} \quad y = -\dfrac{9}{8}$

(b)
$$\frac{dx}{dt} = 6t \quad \frac{dy}{dt} = 3 - 9t^2$$

$$\frac{dy}{dx} = \frac{3 - 9t^2}{6t} = \frac{1 - 3t^2}{2t}$$

$$\frac{dy}{dx} = 0 \text{ for turning points}$$

$$t = \pm \frac{1}{\sqrt{3}}.$$

This helps us find the turning points. The use of the parameters makes the sketching much easier to establish.

(c)
$$x = 3t^2 \Rightarrow t \pm \sqrt{\frac{x}{3}}$$

$$y = 3t - 3t^3 = 3\sqrt{\frac{x}{3}} - 3\frac{x}{3}\sqrt{\frac{x}{3}}$$

$$y = \sqrt{3}\sqrt{x} - x\frac{\sqrt{x}}{\sqrt{3}}$$

$$y = \sqrt{3x} - \frac{x\sqrt{x}\sqrt{3}}{3}.$$

This is rather difficult to sketch or $y = -3\sqrt{\frac{x}{3}} + x\sqrt{\frac{x}{3}}$ when $t = -\sqrt{\frac{x}{3}}$.

4.
$$f(x) = \frac{1}{\sqrt{1+x}} + \sqrt{1-x}$$

$$= (1+x)^{-\frac{1}{2}} + (1-x)^{\frac{1}{2}}$$

$$= 1 + \left(-\frac{1}{2}\right)x + \left(-\frac{1}{2}\right)\left(-\frac{3}{2}\right)\frac{x^2}{2!} + \left(-\frac{1}{2}\right)\left(-\frac{3}{2}\right)\left(-\frac{5}{2}\right)\frac{x^3}{3!}$$

$$+ 1 + \frac{1}{2}(-x) + \frac{1}{2}\left(-\frac{1}{2}\right)\frac{(-x)^2}{2!} + \frac{1}{2}\left(-\frac{1}{2}\right)\left(-\frac{3}{2}\right)\frac{(-x)^3}{3!}$$

$$= 1 - \frac{x}{2} + \frac{3}{8}x^2 - \frac{5}{16}x^3 + 1 - \frac{1}{2}x - \frac{1}{8}x^2 - \frac{1}{16}x^3$$

$$= 2 - x + \frac{1}{4}x^2 - \frac{3}{8}x^3.$$

5.

$$V = \frac{4}{3}\pi r^3 \qquad\qquad S = \frac{dV}{dr} = 4\pi r^2$$

$$\frac{\frac{dV}{dt}}{\frac{dr}{dt}} = 4\pi r^2 \qquad\qquad \frac{dS}{dr} = 8\pi$$

$$\frac{\frac{dS}{dt}}{\frac{dr}{dt}} = 8\pi r \qquad\qquad \frac{dS}{dt} = 8\pi r \frac{dr}{dt}$$

$$\frac{dr}{dt} = \frac{\frac{dS}{dt}}{8\pi r} = \frac{1}{8\pi \, 100}$$

$$\frac{dS}{dt} = 1 \text{ mm}^2/\text{s} \qquad \frac{dV}{dt} = 4\pi \, 100^2 \frac{1}{\pi \, 800}$$

$$= \frac{4 \times 10000}{800} = 50 \text{ mm}^3/\text{s}.$$

6.

$$\overrightarrow{OA} = \mathbf{i} + 2\mathbf{j} + 3\mathbf{k} = \begin{pmatrix} 1 \\ 2 \\ 3 \end{pmatrix}$$

$$\overrightarrow{OB} = 2\mathbf{i} + 2\mathbf{j} + 2\mathbf{k} = \begin{pmatrix} 2 \\ 2 \\ 2 \end{pmatrix}$$

$$\overrightarrow{OC} = 3\mathbf{i} + 5\mathbf{k} = \begin{pmatrix} 3 \\ 0 \\ 5 \end{pmatrix}$$

(i) $|\mathbf{a}| = \sqrt{1^2 + 2^2 + 3^2} = \sqrt{14} = 3.74$ to 3 s.f.

(ii) $|\mathbf{b} - \mathbf{a}| = |2\mathbf{i} + 2\mathbf{j} + 2\mathbf{k} - \mathbf{i} - 2\mathbf{j} - 3\mathbf{k}|$

$$= |\mathbf{i} - \mathbf{k}| = \sqrt{(1)^2 + (-1)^2} = \sqrt{2} = 1.414 \text{ to 3 d.p.}$$

(iii) $|2\mathbf{c} - \frac{1}{2}\mathbf{a}| = |6\mathbf{i} + 10\mathbf{k} - \frac{1}{2}\mathbf{i} - \mathbf{j} - \frac{3}{2}\mathbf{k}|$

$$= \left|\tfrac{11}{2}\mathbf{i} - \mathbf{j} + \tfrac{17}{2}\mathbf{k}\right|$$

$$= \sqrt{\left(\tfrac{11}{2}\right)^2 + (-1)^2 \left(\tfrac{17}{2}\right)^2}$$

$$= \sqrt{\tfrac{121}{4} + 1 + \tfrac{189}{4}}$$

$$= \sqrt{\tfrac{314}{4}} = 8.86 \text{ to 3 s.f.}$$

7. (i) $\displaystyle\int \underset{②}{x} \underset{①}{2^x}\,dx = \frac{1}{\ln 2}2^x x - \int \frac{2^x}{\ln 2}\,dx$

$$= \frac{1}{\ln 2}2^x x - \frac{1}{\ln 2} \times \frac{2^x}{\ln 2} + c$$

$$= \frac{2^x x}{\ln 2} - \frac{1}{(\ln 2)^2}2^x + c$$

(ii) $\displaystyle\int \sin x\, e^x \cos x\,dx = \int \sin x \cos x\, e^x\,dx$

$$= \int \frac{\sin 2x}{2} e^x\,dx = \frac{1}{2}\int \underset{②}{\sin 2x}\, \underset{①}{e^x}\,dx$$

$$= \frac{1}{2}e^x \sin 2x - \int \frac{1}{2}e^x\, 2\cos 2x\,dx$$

$$= \frac{1}{2}e^x \sin 2x - \int \underset{①}{e^x}\, \underset{②}{\cos 2x}\,dx$$

$$= \frac{1}{2}e^x \sin 2x - \left[e^x \cos 2x - \int e^x(-2\sin 2x)\,dx\right]$$

$$= \frac{1}{2}e^x \sin 2x - e^x \cos 2x - 2\int e^x \sin 2x\,dx$$

$\displaystyle\frac{5}{2}\int \sin 2x\, e^x\,dx = \frac{1}{2}e^x \sin 2x - e^x \cos 2x$

$\displaystyle\int \sin x \cos x\, e^x\,dx = \frac{1}{10}e^x \sin 2x - \frac{1}{5}e^x \cos 2x + c$

(iii) $\displaystyle\int \underset{①}{x^2}\, \underset{②}{\ln 2x}\,dx = \frac{x^3}{3}\ln 2x - \int \frac{x^3}{3}\cdot\frac{1}{x}\,dx$

$$= \frac{x^3}{3}\ln 2x - \frac{1}{3} \times \frac{x^3}{3} + c$$

$$= \frac{x^3}{3}\ln 2x - \frac{x^3}{9} + c.$$

TOTAL FOR PAPER: 75 MARKS

GCE Examinations

Test Paper 9

Advanced Level

Core Mathematics C4

Time: 1 hour 30 minutes

Instructions and Information

Candidates may use any calculator allowed by the regulations of their Examination Board.

Full marks are awarded for correct answers to ALL questions.

This paper has seven questions.

You can start working with any question and you must label clearly all parts.

1. $f(x) = \dfrac{x^3 + 2x^2 - 9x - 19}{(x+2)^2(x^2-5)}$.

Given that f(x) can be expressed in the form.

$$f(x) = \dfrac{A}{x+2} + \dfrac{A}{(x+2)^2} - \dfrac{A}{x^2-5}$$

(a) Find the value of A and check the solution. (6)

(b) Hence or otherwise find the series expansion of f(x) in ascending power of x, up to and including the term to x^2. Simplify each term and show that $f(x) \approx 0.95 - 0.5x + 0.3525x^2$. (8)

---

2. Integrate the following:

(i) $\sin^2 x$ (ii) $\cos^2 x$ (iii) $\tan^2 x$ (iv) $\cot^2 x$ (v) $\sec^2 x$ and $\csc^2 x$. (8)

---

3. The cartesian equation of the curve shown in Fig. 1 is given $\dfrac{x^2}{4} + \dfrac{y^2}{9} = 1$.

Verify that the parametric equations are $x = 2\sin t$ and $y = 3\cos t$ for $0 \leq t \leq 2\pi$.

Fig. 1

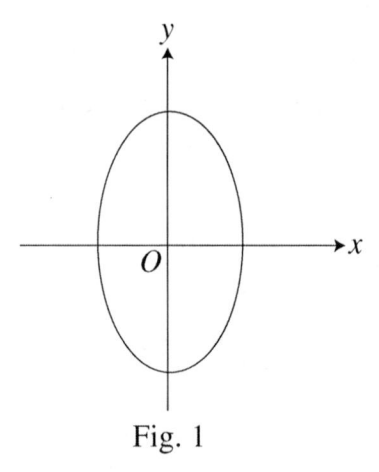

(a) Find an expression for $\dfrac{dy}{dx}$ in terms of the parameter t. (6)

(b) Determine the equations of the tangents at $t = 0$ and $t = \pi$ and sketch these tangents on Fig. 1 (8)

4. Sketch the curve with the equation $y = 5\cos\frac{x}{4}$, $0 \leq x \leq 4\pi$, by considering five ordinates at $x = 0, \pi, 2\pi, 3\pi$ and 4π.

Shade the curve for $0 \leq x \leq 2\pi$.

(a) Find, by integration the area of the shaded region. (4)

This region is rotated through 2π radians about the x-axis.

(b) Find the volume of the solid generated (6)

---

5. (a) Determine the exact value of

$$\int_2^5 x \ln x \, dx.$$

(b)

x	2	3	4	5
y	2 ln 2	3 ln 3	4 ln 4	5 ln 5

Use the trapezium rule to find an approximate value for the definite integral above to 4 significant figures.

Compare the two results obtained in (a) and (b) and state if it is an over estimate or an under estimate. (10)

---

6. The following pairs of lines are given

l_1: $\mathbf{r}_1 = (3\mathbf{i} + \mathbf{j} - 4\mathbf{k}) + \lambda(2\mathbf{i} - 3\mathbf{j} + \mathbf{k})$

l_1: $\mathbf{r}_2 = (-2\mathbf{i} + 4\mathbf{j} + \mathbf{k}) + \mu(-\mathbf{i} + 4\mathbf{j} - 7\mathbf{k})$

determine that the lines are neither parallel nor intersect. (8)

---

7. (a) A horizontal through is 5 m long and has a trapezoidal cross sectional area as shown in Fig. 2

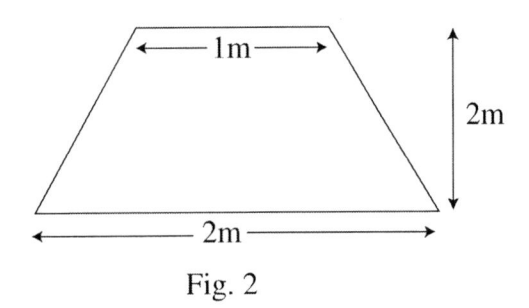

Fig. 2

Water runs into the trough at the rate of 1×10^{-3} m³ per second.

Find the rate at which the water is rising when the height of the water is 1 m. **(6)**

(b) Find the equation of the normal at a point θ on the ellipse

$x = 2\cos\theta$ and $y = 3\sin\theta$. **(5)**

TOTAL FOR PAPER: 75 MARKS

GCE Examinations

Test Paper 9 Solutions

Advanced Level

Core Mathematics C4

1. $f(x) \equiv \dfrac{x^3 + 2x^2 - 9x - 19}{(x+2)^2(x^2-5)} \equiv \dfrac{A}{x+2} + \dfrac{A}{(x+2)^2} - \dfrac{A}{x^2-5}$

$\equiv \dfrac{A(x+2)(x^2-5) + A(x^2-5) - A(x+2)^2}{(x+2)^2(x^2-5)}$

$x^3 + 2x^2 - 9x - 19 = A(x+2)(x^2-5) + A(x^2-5) - A(x+2)^2$

if $x = -2$

$-8 + 8 + 18 - 19 = -A \Rightarrow \boxed{A = 1}$

if $x = 0$

$-19 = -10 - 5 - 4A \Rightarrow \boxed{A = 1}$

if $x = 1$

$1 + 2 - 9 - 19 = -12A - 4A - 9A \Rightarrow A = 1$

$\therefore f(x) = \dfrac{1}{x+2} + \dfrac{1}{(x+2)^2} - \dfrac{1}{x^2-5}$

$= \dfrac{(x+2)(x^2-5) + (x^2-5) - (x+2)^2}{(x+2)^2(x^2-5)}$

$= \dfrac{x^3 + 2x^2 - 5x - 10 + x^2 - 5 - x^2 - 4x - 4}{(x+2)^2(x^2-5)} = \dfrac{x^3 + 2x^2 - 9x - 19}{(x+2)^2(x^2-5)}.$

(b) $f(x) = (x+2)^{-1} + (x+2)^{-2} - (x^2-5)^{-1}$

$= 2^{-1}(1 + \tfrac{x}{2})^{-1} + 2^{-2}(1 + \tfrac{x}{2})^{-2} - (-5)^{-1}(1 - \tfrac{x^2}{5})^{-1}$

$= \dfrac{1}{2}\left[1 + (-1)\left(\tfrac{x}{2}\right) + \tfrac{(-1)(-2)}{2!}\left(\tfrac{x}{2}\right)^2\right] + \dfrac{1}{4}\left[1 + (-2)\left(\tfrac{x}{2}\right) + \tfrac{(-2)(-3)}{2!}\left(\tfrac{x}{2}\right)^2\right]$

$+ \dfrac{1}{5}\left[1 + (-1)\left(-\tfrac{x^2}{5}\right) + \tfrac{(-1)(-2)}{2!}\left(-\tfrac{x^2}{5}\right)^2\right]$

$= \dfrac{1}{2}\left(1 - \tfrac{x}{2} + \tfrac{x^2}{4}\right) + \dfrac{1}{4}\left(1 - x + 3\tfrac{x^2}{4}\right) + \dfrac{1}{5}\left(1 + \tfrac{x^2}{5}\right)$

$= \dfrac{1}{2} + \dfrac{1}{4} + \dfrac{1}{5} - \dfrac{x}{4} - \dfrac{x}{4} + \dfrac{x^2}{8} + \dfrac{3}{16}x^2 + \dfrac{x^2}{25}$

$= 0.95 - 0.5x + 0.3525x^2.$

2. (i) $\int \sin^2 x \, dx = \int \dfrac{1 - \cos 2x}{2} \, dx = \dfrac{1}{2}x - \dfrac{\sin 2x}{4} + c$

$\cos 2x = 2\cos^2 x - 1 = 1 - 2\sin^2 x$

$\therefore \sin^2 x = \dfrac{1 - \cos 2x}{2}$

$\cos^2 x = \dfrac{\cos 2x + 1}{2}$

(ii) $\int \cos^x dx = \int \dfrac{\cos 2x + 1}{2} \, dx = \dfrac{\sin 2x}{4} + \dfrac{1}{2}x + c$

(iii) $\int \tan^2 x \, dx = \int (\sec^2 x - 1) \, dx = \tan x - x + c$

(iv) $\int \cot^2 dx = \int (\operatorname{cosec}^2 x - 1) \, dx = -\cot x - x + c$

$1 + \tan^2 x = \sec^2 x \qquad 1 + \cot^2 x = \operatorname{cosec}^2 x$

(v) $\int \sec^2 x \, dx = \tan x + c$

(vi) $\int \operatorname{cosec}^2 x \, dx = -\cot x + c.$

---

3. $\qquad \dfrac{x^2}{4} + \dfrac{y^2}{9} = 1 \quad \cdots (1)$

Substituting $x = 2 \sin t$ and $y = 3 \cos t$ in equation (1)

$\dfrac{(2 \sin t)^2}{4} + \dfrac{(3 \cos t)^2}{9} = \dfrac{4 \sin^2 t}{4} + \dfrac{9 \cos^2 t}{9} = \sin^2 t + \cos^2 = 1$

(a) $x = 2 \sin t \qquad \dfrac{dx}{dt} = 2 \cos t \qquad \cdots (2)$

$y = 3 \cos t \qquad \dfrac{dy}{dt} = -3 \sin t \qquad \cdots (3)$

$\dfrac{dy}{dx} = \dfrac{\frac{dy}{dt}}{\frac{dx}{dt}} = -\dfrac{3 \sin t}{2 \cos t} = -\dfrac{3}{2} \tan t$

(b) $m = \dfrac{dy}{dx} = -\dfrac{3}{2}\tan t$ when $t = 0$

$m = 0$, $y = mx + c \Rightarrow y = c = 3\cos 0 = 3 \Rightarrow \boxed{y = 3}$

the equation of the tangent at $t = 0$

$m = \dfrac{dy}{dx} = -\dfrac{3}{2}\tan t$ when $t = 0$

$m = -\dfrac{3}{2}\tan \pi = 0$, $y = mx + c \Rightarrow y = c = 3\cos \pi = -3 \Rightarrow \boxed{y = -3}$

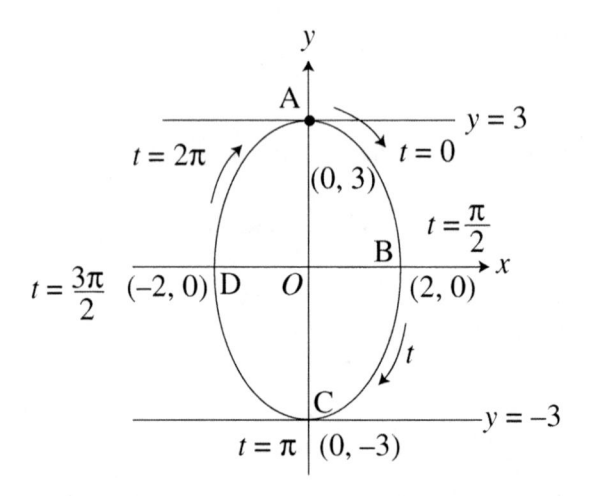

---

4. $y = 5\cos\dfrac{x}{4}$

x	0	π	2π	3π	4π
y	5	$\dfrac{5}{\sqrt{2}}$	0	$-\dfrac{5}{\sqrt{2}}$	-5

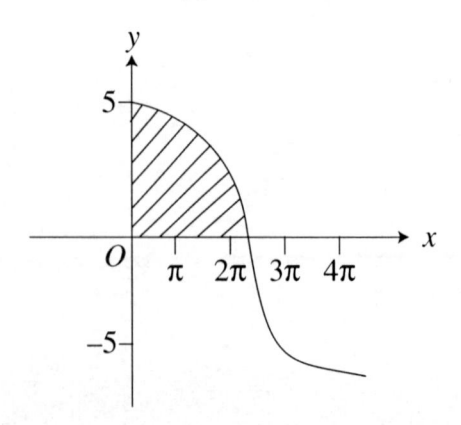

(a) $\int_0^{2\pi} 5\cos\frac{x}{4}\,dx = \left[20\sin\frac{x}{4}\right]_0^{2\pi} = 20\sin\frac{2\pi}{4} - 20\sin 0 = 20$ s.u.

(b) $\pi\int_0^{2\pi} y^2\,dx = \pi\int_0^{2\pi} 25\cos^2\frac{x}{4}\,dx = \pi$

$\cos\frac{x}{2} = 2\cos^2\frac{x}{4} - 1 \Rightarrow \cos^2\frac{x}{4} = \frac{1}{2}\left(\cos\frac{x}{2} + 1\right)$

$\pi\int_0^{2\pi} 25\cos^2\frac{x}{4}\,dx = 25\pi\int_0^{2\pi}\frac{1}{2}\left(\cos\frac{x}{2} + 1\right)dx$

$= \frac{25}{2}\pi\left[\frac{\sin\frac{x}{2}}{\frac{1}{2}} + x\right]_0^{2\pi} = 25\pi(\sin\pi) + \frac{25}{2}\pi\,2\pi = 25\pi^2$ cubic units.

---

5. (a) $\int_2^5 x\ln x\,dx = \left[\frac{x^2}{2}\ln x\right]_2^5 - \int_2^5 \frac{x^2}{2}\times\frac{1}{x}\,dx$
       ① ②

$= \frac{25}{2}\ln 5 - 2\ln 2 - \frac{1}{2}\left[\frac{x^2}{2}\right]_2^5$

$= 12.5\ln 5 - 2\ln 2 - \frac{25}{4} + \frac{1}{2}\times 2$

$= 12.5\ln 5 - 2\ln 2 - \frac{25}{4} + \frac{4}{4}$

$= 12.5\ln 5 - 2\ln 2 - \frac{21}{4} = \ln\frac{5^{12.5}}{4} - \frac{21}{4}.$

(b) $\int_2^5 x\ln x\,dx \approx \frac{1}{2}\left[2\ln 2 + 5\ln 5 + 2(3\ln 3 + 4\ln 4)\right]$

$\approx \frac{1}{2}\left[\ln 4\times 5^5 \times 3^6 \times 4^8\right]$

$= \frac{1}{2}\ln 5971968\times 10^{11} = 13.55775627 = 13.56$ to 4 s.f.

$\ln\frac{5^{12.5}}{4} - \frac{21}{4} = 13.48167954.$

The overestimate is $13.55775627 - 13.48167954 = 0.07607673$ square units.

6. Examine the direction ratios of the direction vectors.

The direction ratios of l_1 are $2: -3: 1$.

The direction ratios of l_2 are $-1: 4: -7$.

These direction ratios are not equal and therefore the lines are <u>not</u> parallel.

If the lines intersect, then they have a common point.

Equating the coefficients of **i, j, k** we have

$3 + 2\lambda = -2 - \mu$ or $2\lambda = -5 - \mu \cdots (1)$

$1 - 3\lambda = 4 + 4\mu$ or $3\lambda = -3 - 4\mu \cdots (2)$

$-4 + \lambda = 1 - 7\mu$ or $\lambda = 5 - 7\mu \cdots (3)$

Solving equations (1) and (3) in order to find the values of λ and μ

$2(5 - 7\mu) = -5 - \mu$ or $10 - 14\mu = -5 - \mu$ or $\mu = \frac{15}{13}$ and substituting in (3)

$\lambda = 5 - \frac{105}{13} = -\frac{40}{3}.$

If the lines intersect then equation (2) is true for these values, but equation (2) is not verified and the lines are skew.

---

7. (a) $V = $ volume of the trough

$\qquad = $ area of trapezium $\times$ length of trough

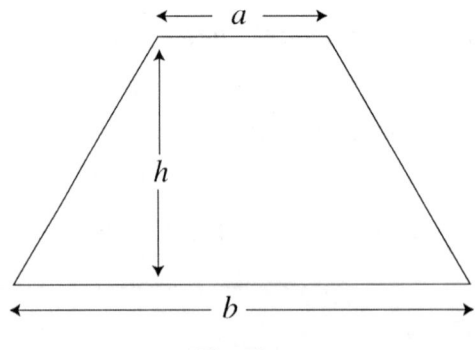

Fig. 2

$$V = \frac{1}{2}(a+b)hl = \frac{1}{2}(1+2)5h = \frac{15}{2}h$$

$$\frac{dV}{dh} = \frac{15}{2} = \frac{\frac{dV}{dt}}{\frac{dh}{dt}} = 7.5$$

$$\frac{dh}{dt} = \frac{dV}{dt} \times \frac{1}{7.5} = \frac{1 \times 10^{-3}}{7.5} \text{ ms}^{-1}$$

(b) $x = 2\cos\theta \qquad \dfrac{d}{d\theta} = -2\sin\theta$

$\quad y = 3\sin\theta \qquad \dfrac{dy}{d\theta} = 3\cos\theta$

$$\frac{dy}{dx} = \frac{\frac{dy}{d\theta}}{\frac{dx}{d\theta}} = \frac{3\cos\theta}{-2\sin\theta} = -\frac{3}{2}\cot\theta$$

$y = m_1 x + c \quad$ where $m_2 = \dfrac{2}{3}\tan\theta$

the gradient of the normal since $m_1 m_2 = -1$

$\therefore y = m_2 x + c$

$3\sin\theta = \dfrac{2}{3}\tan\theta(2\cos\theta) + c$

$c = 3\sin\theta - \dfrac{4}{3}\tan\theta\cos\theta = 3\sin\theta - \dfrac{4}{3}\sin\theta = \dfrac{5}{3}\sin\theta$

$$\boxed{y = \left(\frac{2}{3}\tan\theta\right)x + \frac{5}{3}\sin\theta}$$

TOTAL FOR PAPER: 75 MARKS

GCE Examinations

Test Paper 10

Advanced Level

Core Mathematics C4

Time: 1 hour 30 minutes

Instructions and Information

Candidates may use any calculator allowed by the regulations of their Examination Board.

Full marks are awarded for correct answers to ALL questions.

This paper has seven questions.

You can start working with any question and you must label clearly all parts.

1. (a) Express the partial fractions
$$\frac{1}{x+1} - \frac{2}{(x+1)^2} + \frac{3}{x^2+1} = f(x)$$
   into a single fraction and show that $f(x) = \dfrac{x^3 + 2x^2 + 7x + 2}{(x+1)^2(x^2+1)}$. (4)

   (b) Hence decompose the last expression of f(x) into partial fractions and show that it is equal to $\dfrac{1}{x+1} - \dfrac{2}{(x+1)^2} + \dfrac{3}{x^2+1}$. (7)

2. (a) $\displaystyle\int e^{-qx} \cos px \, qx$. (8)

   (b) $\displaystyle\int (x+1) \ln \sqrt{x+1} \, dx$. (4)

   (c) $\displaystyle\int \sin^3 x \, dx$. (4)

3. The cartesian equation of the circle C is $x^2 - 8x + y^2 + 6x = 0$

   (a) Find the coordinates of the centre of C and the radius of $C = 0$. (4)

   (b) Sketch the curve C. (2)

   (c) Determine the parametric equations for C. (4)

   (d) Show that the gradient of the circle at p(x, y) is $\dfrac{dy}{dx} = \dfrac{4-x}{y+3}$

   hence find the cartesian equations of the tangents at (0, 0) and (0, −6)
   and show that they are at right angles and they intersect at $Q(-\frac{9}{4}, -3)$. (7)

4. Write down the first five terms in the expansion $\sqrt{1 - 2x}$. (4)

5. Evaluate approximately to three decimal places using the trapezoital rule, the definite integral
$$\int_0^{\frac{\pi}{4}} \sqrt{\sin x} \, dx$$
using 11 ordinates, finding first the missing results. (5)

x	0	$\frac{\pi}{40}$	$\frac{2\pi}{40}$	$\frac{3\pi}{40}$	$\frac{4\pi}{40}$	$\frac{5\pi}{40}$
$\sqrt{\sin x}$	0		0.3955		0.5559	

x	$\frac{6\pi}{40}$	$\frac{7\pi}{40}$	$\frac{8\pi}{40}$	$\frac{9\pi}{40}$	$\frac{10\pi}{40}$
$\sqrt{\sin x}$		0.7228		0.8059.	

(3)

---

6. The vector equations of the three lines are given below:

$$l_1 : \mathbf{r} = 2\mathbf{i} + 3\mathbf{j} + 4\mathbf{k} + \lambda(2\mathbf{i} - 3\mathbf{j} + 5\mathbf{k})$$
$$l_2 : \mathbf{r} = 2\mathbf{i} - 5\mathbf{j} + 4\mathbf{k} + \mu(4\mathbf{i} - 6\mathbf{j} + 10\mathbf{k})$$
$$l_3 : \mathbf{r} = -\mathbf{i} + \mathbf{j} + 3\mathbf{k} + t(\mathbf{i} - \mathbf{j} + 2\mathbf{k})$$

Find which pair of lines

(i) are parallel to each other (2)

(ii) are skew (4)

(iii) intersect with other. (4)

---

7. Solve the differential equation $(1 + x^4)\,dy - x^3 y\,dx = 0$

given that $y = 3$ when $x = 1$. (9)

TOTAL FOR PAPER: 75 MARKS

GCE Examinations

Test Paper 10 Solutions

Advanced Level

Core Mathematics C4

1. (a)
$$\frac{1}{x+1} - \frac{2}{(x+1)^2} + \frac{3}{x^2+1} = f(x)$$

$$= \frac{(x+1)(x^2+1)}{(x+1)^2(x^2+1)} - \frac{2(x^2+1)}{(x+1)^2(x^2+1)} + \frac{3(x+1)^2}{(x+1)^2(x^2+1)}$$

$$= \frac{x^3 + x^2 + x + 1 - 2x^2 - 2 + 3x^2 + 6x + 3}{(x+1)^2(x^2+1)}$$

$$= \frac{x^3 + 2x^2 + 7x + 2}{(x+1)^2(x^2+1)}.$$

(b)
$$\frac{x^3 + 2x^2 + 7x + 2}{(x+1)^2(x^2+1)} \equiv \frac{A}{x+1} + \frac{B}{(x+1)^2} + \frac{Cx+D}{x^2+1}$$

$$x^3 + 2x^2 + 7x + 2 \equiv A(x+1)(x^2+1) + B(x^2+1) + (Cx+D)(x+1)^2.$$

Using the cover up rule, we have:

if $x = -1$, $\quad -1 + 2 - 7 + 2 = 2B = -4 \Rightarrow B = -2$

if $x = 0$, $\quad 2 = A + B + D = A - 2 + D \Rightarrow \boxed{A + D = 4}$

if $x = 2$, $\quad 8 + 8 + 14 + 2 = 3A(5) + (-2)(5) + (2C+D)9$

$$32 = 15A - 10 + 18C + 9D$$

$$\therefore 15A + 18C + 9D = 42 \quad \ldots (1)$$

if $x = 1$ $\quad 1 + 2 + 7 + 2 = 2A2 - 2(2) + (C+D)4$

$$4A + 4C + 4D = 16$$

$$A + C + D = 4$$

$$\boxed{C = 0}$$

From (1) $15A + 9D = 42$

$$15A + 9D = 42 \quad \ldots (1)$$
$$\underline{A + D = 4} \quad \ldots (2) \times -15$$

$$15A + 9D = 42$$
$$\underline{-15A - 15D = -60}$$
$$-6D = -18$$
$$\boxed{D = 3}$$

$$A + D = 4 \Rightarrow \boxed{A = 1}$$

$$\therefore f(x) = \frac{1}{x+1} - \frac{2}{(x+1)^2} + \frac{3}{x^2+1}.$$

---

2. (a)

$$\int \underset{\textcircled{2}}{e^{-qx}} \underset{\textcircled{1}}{\cos px} \, qx = \frac{\sin px}{p} e^{-qx} - \int \frac{\sin px}{p} (-qe^{-qx}) \, qx$$

$$= \frac{e^{-qx}}{p} \sin px + \frac{q}{p} \int \underset{\textcircled{1}}{\sin px} \, \underset{\textcircled{2}}{e^{-qx}} \, qx$$

$$= \frac{e^{-qx}}{p} \sin px + \frac{q}{p} \left[-\frac{\cos px}{p} e^{-qx} - \right.$$

$$\left. \int -\frac{\cos px}{p} (-qe^{-qx}) \, qx \right]$$

$$= \frac{1}{p} e^{-qx} \sin px - \frac{q}{p^2} e^{-qx} \cos px - \frac{q^2}{p^2} \int e^{-px} \cos px \, qx$$

$$= \left(1 + \frac{q^2}{p^2}\right) \int e^{-qx} \cos px \, qx = \frac{1}{p} e^{-qx} \sin px - \frac{d}{p^2} e^{-qx} \cos px$$

$$\therefore \int e^{-qx} \cos px \, qx = \frac{p^2}{p^2 + q^2} \left(\frac{1}{p} e^{-qx} \sin px - \frac{q}{p^2} e^{-qx} \cos px \right)$$

$$= \frac{e^{-qx}}{p^2 + q^2} (p \sin px - q \cos px).$$

(b) $\int (x+1) \ln \sqrt{x+1}\, dx = \dfrac{(x+1)^2}{2} \ln \sqrt{x+1}$
　　　　①　　　②

$$-\int \dfrac{(x+1)^2}{2} \dfrac{1}{\sqrt{(x+1)^2}} \times \dfrac{1}{2}(x+1)^{-\tfrac{1}{2}}\, dx$$

$$= \dfrac{(x+1)^2}{2} \ln \sqrt{x+1} - \dfrac{1}{4}\int \dfrac{(x+1)^2}{x+1}\, dx$$

$$= \dfrac{(x+1)^2}{2} \ln \sqrt{x+1} - \dfrac{1}{4}\dfrac{(x+1)^2}{2} + c$$

(c) $\int \sin^3 x\, dx = \int \sin^2 x \, \sin x\, dx$

$$= -\int (1 - \cos^2 x)\, d(\cos x)$$

$$= -\cos x + \dfrac{\cos^3 x}{3} + c.$$

---

3. (a) $x^2 - 8x + y^2 + 6y = 0$ using the method of completing the square

$$(x-4)^2 - 16 + (y+3)^2 - 9 = 0$$
$$(x-4)^2 + (y+3)^2 = 25 = 5^2$$
$$C(4, -3) \quad r = 5.$$

(b)

(c) $$x = 5\sin t + 4$$
$$y = 5\cos t - 3$$
$$(5\sin t + 4 - 4)^2 + (5\cos t - 3 + 3)^2 = 25$$
$$25\sin^2 t + 25\cos^2 t = 25$$
$$\sin^2 t + \cos^2 t = 1$$

(d) $x^2 - 8x + y^2 + 6x = 0$ differentiating with respect to x
$$2x - 8 + 2y\frac{dy}{dx} + 6\frac{dy}{dx} = 0$$
$$\frac{dy}{dx} = \frac{8 - 2x}{2y + 6} = \frac{4 - x}{y + 3}$$
$$\frac{dy}{dx} = m = \frac{4}{3} \quad \boxed{y = \frac{4}{3}x}$$
$$\frac{dy}{dx} = \frac{4 - 0}{-6 + 3} = -\frac{4}{3} \quad y = -\frac{4}{3}x + c$$
$$-6 = c \quad \boxed{y = -\frac{4}{3}x - 6}$$

$$\frac{4}{3}x = -\frac{4}{3}x = -6$$
$$\frac{8}{3}x = -6 \Rightarrow x = -\frac{18}{8} = -\frac{9}{4}$$
$$Q\left(-\frac{9}{4}, -3\right) \quad y = \frac{4}{3}\left(-\frac{9}{4}\right) = -3,$$

---

4. $$\sqrt{1 - 2x} = (1 - 2x)^{\frac{1}{2}} = 1 + \frac{1}{2}(-2x) + \frac{1}{2}\left(-\frac{1}{2}\right)\frac{(-2x)^2}{2!}$$
$$+ \frac{1}{2}\left(-\frac{1}{2}\right)\left(-\frac{3}{2}\right)\frac{(-2x)^3}{3!}$$
$$+ \frac{1}{2}\left(-\frac{1}{2}\right)\left(-\frac{3}{2}\right)\left(-\frac{5}{2}\right)\frac{(-2x)^4}{4!}$$
$$= 1 - x - \frac{1}{2}x^2 - \frac{1}{2}x^3 - \frac{5}{8}x^4.$$

5.

x	0	$\frac{\pi}{40}$	$\frac{2\pi}{40}$	$\frac{3\pi}{40}$	$\frac{4\pi}{40}$	$\frac{5\pi}{40}$	$\frac{6\pi}{40}$	$\frac{7\pi}{40}$	$\frac{8\pi}{40}$	$\frac{9\pi}{40}$	$\frac{10\pi}{40}$
$\sqrt{\sin x}$	0	0.7801	0.3955	0.4832	0.5559	0.6186	0.6738	0.7278	0.8054	0.8409	

$$\int_0^{\frac{\pi}{4}} \sqrt{\sin x}\, dx \approx \frac{\pi}{2 \times 40}\left[0+0.8409+2(0.2801+0.3955+0.4832+0.5559+0.6186\right.$$
$$\left.+0.6738+0.7228+0.7667+0.8059)\right]$$
$$= 0.449$$

where $h = \frac{\pi}{4 \times 10} = \frac{\pi}{40}$.

---

6. (i) l_1 and l_2 are parallel

since the direction ratios are the same $2: -3: 5$ and $4: -6: 10$

(iii)
$$l_1: \mathbf{r} = \begin{pmatrix} 2 \\ 3 \\ 4 \end{pmatrix} + \lambda \begin{pmatrix} 2 \\ -3 \\ 5 \end{pmatrix} = \begin{pmatrix} x \\ y \\ z \end{pmatrix}$$

$$l_3: \mathbf{r} = \begin{pmatrix} -1 \\ 1 \\ 3 \end{pmatrix} + t \begin{pmatrix} 1 \\ -1 \\ 2 \end{pmatrix} = \begin{pmatrix} x \\ y \\ z \end{pmatrix}$$

for x $\quad 2 + 2\lambda = -1 + t \Rightarrow 2\lambda - t = -3 \quad \ldots (1)$

y $\quad 3 - 3\lambda = 1 - t \Rightarrow -3\lambda + t = -2 \quad \ldots (2)$

z $\quad 4 + 5\lambda = 3 + 2t \Rightarrow 5\lambda - 2t = -1 \quad \ldots (3)$

$(1) + (2) \quad -\lambda = -5 \Rightarrow \boxed{\lambda = 5}$

substituting in $(1) -t = -3 - 10 \quad \boxed{t = 13}$

verify $\lambda = 5$ and $t = 13$ in (3)

$25 - 26 = -1$

l_1 and l_3 intersect

(ii)
$$l_2: \mathbf{r} = \begin{pmatrix} 2 \\ -5 \\ 1 \end{pmatrix} + \mu \begin{pmatrix} 4 \\ -6 \\ 10 \end{pmatrix}$$

$$l_3: \mathbf{r} = \begin{pmatrix} -1 \\ 1 \\ 3 \end{pmatrix} + t \begin{pmatrix} 1 \\ -1 \\ 2 \end{pmatrix}$$

For x $2 + 4\mu = -1 + t \Rightarrow 4\mu - t = 3$ ... (1)

y $-5 - 6\mu = 1 - t \Rightarrow 6\mu + t = -6$ ... (2)

z $1 + 10\mu = 3 + 2t \Rightarrow 10\mu - 2t = 2$ ... (3)

$$(1) - (2) \quad -2\mu = 3 \Rightarrow \boxed{\mu = -\frac{3}{2}}$$

substituting in (1) $-t = -3 - 4\left(-\frac{3}{2}\right)$ $\boxed{t = -3}$

verify (3) for $t = -3$ and $\mu = -\frac{3}{2}$

$$1 + 10\left(-\frac{3}{2}\right) = 3 + 2(-3)$$

$$1 - 15 \neq -3$$

l_2 and l_3 are inconsistent for these values of μ and t and therefore are skew lines.

---

7. $(1 + x^4)\, dy = x^3 y\, dx$

$$\int \frac{dy}{y} = \int \frac{x^3}{1+x^4}\, dx$$

$$\ln y = \frac{1}{4} \int \frac{d(1+x^4)}{1+x^4} = \frac{1}{4} \ln(1+x^4) + \ln A$$

$$\ln y = \ln(1+x^4)^{\frac{1}{4}} A$$

$$y = A(1+x^4)^{\frac{1}{4}}$$

$$3 = A(1+1)^{\frac{1}{4}}$$

$$A = \frac{3}{\sqrt[4]{2}} = 2.52$$

$$\therefore y = 2.52(1+x^4)^{\frac{1}{4}}.$$

TOTAL FOR PAPER: 75 MARKS